Beyond the Nanoworld

Nanoworld

Quarks, Leptons, and Gauge Bosons

Hans Günter Dosch

A K Peters, Ltd.
Wellesley, Massachusetts

Editorial, Sales, and Customer Service Office

A K Peters, Ltd.
888 Worcester Street, Suite 230
Wellesley, MA 02482
www.akpeters.com

Library of Congress Cataloging-in-Publication Data

Dosch, Hans Günter.
   [Jenseits der Nanowelt. English]
   Beyond the nanoworld : quarks, leptons, and gauge bosons / Hans Günter Dosch.
     p. cm.
   Includes index.
   ISBN 978-1-56881-345-5 (alk. paper)
   1. Particles (Nuclear physics) --History-- 20th century.   I. Title.

   QC793.16D6713 2007
   539.7'20904--dc22

                             2007041605

Front cover images, left to right: ALICE muon spectrometer; ATLAS cavern; TCP Time Projection Chamber; ALICE muon spectrometer. ALICE Collaboration, CERN Document Server. ©CERN.

Back cover image: Gluon discovery in 1979. DESY, Hamburg, Germany.

Printed in the United States of America
12 11 10 09 08   10 9 8 7 6 5 4 3 2 1

In memory of *J. H. D. Jensen*
For *Simon, Linus, Jonas, and Philipp*

Figure Credits

Figures 1.2, 2.10, 2.11, and 2.12 reprinted with permission from: H. L. Anderson. "Early History of Physics with Accelerators." *Journal de Physique*, 43:C8 (1982), C8-101.

Figures 2.1, 2.2, 2.3, and 2.4 reprinted with permission from: Ch. Peyrou. "The Role of Cosmic Rays in the Development of Particle Physics." *Journal de Physique*, 43:C8 (1982), C8-7.

Figure 1.25 reprinted with permission from: C. D. Anderson. "The Positive Electron." *Phys. Rev.*, 43 (1933), 491. ©1933 The American Physical Society. http://link.aps.org/abstract/PR/v43/p491.[*]

Figure 1.27 reprinted with permission from: S. H. Neddermeyer and C. D. Anderson. "Note on the Nature of Cosmic Ray Particles." *Phys. Rev.*, 51 (1937), 884. ©1937 The American Physical Society. http://link.aps.org/abstract/PR/v51/p884.[*]

Figures 3.5, 7.5, and 8.2 reprinted with permission from: S. Donnachie, G. Dosch, P. Landshoff, and O. Nachtmann. *Pomeron Physics and QCD*. Cambridge University Press, Cambridge, UK, 2002.

Figure 4.5 reprinted with permission from: V. E. Barnes et al. "Observation of a Hyperon with Strangeness Minus Three." *Phys. Rev. Lett.*, 12 (1964), 204. ©1964 The American Physical Society. http://link.aps.org/abstract/PRL/v12/p204.[*]

Figure 5.9 reprinted with permission from: R. Schwitters. "Development of Large Detectors for Colliding-Beam Experiments." In *The Rise of the Standard Model*, edited by L. Hoddeson et al., p. 299. Cambridge University Press, Cambridge, UK, 1997.

Figures 6.12, 6.13, and 8.1 reprinted with permission from: W.-M. Yao et al. [Particle Data Group Collaboration]. "Review of Particle Physics." *J. Phys. G.*, 33 (2006), 1.

Contents

Preface

We are now well aware that matter is made of atoms and molecules. Pictures made with scanning–tunneling microscopes make the fundamentally granular structure of matter directly visible. The world of atoms and molecules (which is about a million times smaller than the structures we can see and feel) is also known as the nanoworld. It has found a growing number of applications in technology and is rapidly gaining in importance.

This books deals with the structures we find when we look beyond the nanoworld, namely leptons, quarks and gauge bosons. Leptons and quarks are the smallest, elementary building blocks underlying atomic structure and the gauge bosons mediate the forces between these particles, binding the inner world together. In order to investigate the properties of these particles one needs microscopes with a resolution power that is over a million times higher than that of tunneling microscopes. The huge accelerators that have been built since the middle of the last century can be viewed as such microscopes. It was with the development of these accelerators that the United States began to dominate fundamental research. The first modern accelerators were developed and constructed by E. O. Lawrence in Berkeley in the early 1930s. After World War II, Europe combined its resources; with the establishment of the European Organization for Nuclear Research (CERN) in the 1950s it caught up with the big national laboratories in the US.

Today, the "particle war," as this friendly and scientifically beneficial competition is sometimes called, is focused on discovering the so-called Higgs boson. This is an elementary particle whose existence is an essential feature of an extremely successful theory: the standard model of particle physics. At the moment, it is an open question whether this particle will first be observed at the large Tevatron accelerator at the Fermi National Accelerator Laboratory (Fermilab) in Batavia, Illinois, or at the yet to be completed LHC (Large Hadron Collider) in Geneva, Switzerland—if it will be detected at all.

This book proceeds in roughly historical order, but it is not primarily intended as a history of particle physics. Controversial questions of priority, for example, are not addressed. Rather, I trace the historical development of the concepts of particle physics in order to make them more transparent and intelligible. I also wish to show how closely intertwined theoretical and experimental advances were, and how they mutually influenced each other. I have tried to treat broadly some truly fundamental concepts and treat other fields briefly (even if they played an important role in the development of particle physics, especially if these fields are difficult to describe and if they were superseded by later developments). Of course, my personal opinions had some influence on the content of the book. I hope that if a specialist in particle physics has a look at it he will find nothing that he considers to be wrong, although I am sure that he will think that at least one field has received far too little space (the subfield in which he works).

This book is addressed to readers with a general interest in science, and I have tried to present the subject in as easily intelligible a manner as possible. Modern physics has shown how the human intellect can surmount the limits of every-day experience and how it can explain what we see in science through mind-boggling technological extensions of our human senses. Without mathematics this exploration would be impossible, and mathematics is indispensable if one wants to understand fully the detailed features of particle physics. Nevertheless, I present hardly any formulae, but try to indicate complex mathematical relations in words. This may on occasion lead to distortions, especially in Sections 1.4 and 1.5. An appendix with (relatively simple) formulae can be found at the book's webpage: http://www.thphys.uni-heidelberg.de/~dosch/beyondnano.

The beautiful book *Inward Bound*, by Abraham Pais, who was himself a principle player in the field of particle physics, was for me a valuable guide through the complex development of particle physics; it treats the period from about 1890 to 1990. Some details I owe to reviews on the history of particle physics that were authored by active physicists, but my principal source material is the original literature, published in scientific journals. Beyond the sources for adopted figures, I cite only a few books and review articles; an extensive bibliography can be found on the book's webpage.

I am gratefully obliged to many colleagues for numerous discussions and valuable hints, I want to acknowledge especially the helpful comments and amendments of W. Beiglböck, D. Haidt, B. Lohff, and E. Schücking. I also want to thank Alice and Klaus Peters for their encouragement and help in preparing this American edition.

Heidelberg, October 2007   *Hans Günter Dosch*

1

The Heroic Time

The foundations of modern physics—in contrast to those of classical physics—were laid in the first half of the twentieth century. In theoretical physics, the decisive developments were the establishment of relativity and quantum physics, while experimental physics contributed a secure knowledge of the atomic structure of matter as well as the internal structure of atoms. Nearly all fields of modern science, from astrophysics to medical diagnostics, have been strongly influenced by the methods and concepts developed during this period.

1.1 Introduction

The study of the science of elementary particles goes back to ancient times. In ancient Greek natural philosophy, we already find the hope that by exploring the fundamental constituents of matter, one might reduce sensory experiences —such as sweet and bitter, warm and cold—to these basic components. The idea of using the certainty of mathematics to this end is also traceable to that era. More than 2,000 years ago, this was attempted by Plato, who described his theory of the world in his dialogue *Timaeus*. He made use of the mathematical result that there are only five regular solid figures. He assigned four of these Platonic solids to the four elements of ancient science (see Figure 1.1). To earth he assigned the cube, since "the earth is of the four kinds the most immovable and under the bodies the most plastic." To fire he assigned the tetrahedron, since it has the "least areas and is in all directions the most movable, the most acute and the sharpest." Using similar reasoning, he assigned the icosahedron to water and the octahedron to air. He assigned the remaining dodecahedron,

| Earth | Water | Air | Fire | The Universe |

Figure 1.1. The assignment of the five Platonic solids to the four natural elements and to the universe as a whole.

which is the most similar to the sphere, to the All (the universe). Plato is scientifically very restrained in presenting his theory, indeed more cautious than Galileo would be almost 2,000 years later. Plato claims only that his theory is plausible, not that it is certain.

In his book *The Part and the Whole*, Werner Heisenberg writes that as a college student he found Plato's theory very speculative, but more convincing than the atoms of his schoolbooks with their hooks and eyes.

Ancient atomism was described by Lucretius in the six books of his poem *On the Nature of Things*. Of course, there are essential differences between ancient natural philosophy and modern particle physics, but they have some fundamental ideas, and even some methodological approaches, in common. This is apparent when one looks at the great influence of ancient ideas on modern thought. Plato's influence on Galileo and Kepler is well known. The Scottish physicist Lord Kelvin proposed a very abstract and strikingly modern topological model of the atom as a vortex in the ether in 1886, not long after Hermann Helmholtz had shown that such vortices were stable in an ideal liquid. Therefore, to Lord Kelvin these vortices seemed to furnish a very plausible model for—as he writes—Lucretian atoms.

The first theorists of elementary particles in the modern sense of the term were chemists. In chemistry, the fundamental particles are the constituents of the basic chemical elements. They were called atoms, since they were believed to be indivisible. Important quantitative conclusions were drawn from that atomic theory. With the discovery of the electron, however, it became clear that there are particles that are even simpler than the atoms studied by chemists.

I do not intend to describe the complete history of atomism; I will start with the discovery of the electron at the end of the nineteenth century (1899). From that time on, the name *elementary particle* has been used for a host of different objects. Until 1935, the study of particle physics was rather uncomplicated. The well-established elementary particles were the *proton* and the *neutron*, which were the components of the atomic nucleus, the *electron*, which formed the atomic cloud around the nucleus, and the *photon*, which was the quantum of electromagnetic radiation. In the late 1930s the *mesotron* and the *neutrino* joined the club. The mesotron was introduced in order to explain nuclear forces, while the existence of the neutrino was postulated in order to explain certain types of radioactive decay.

From 1950 onward the situation became more complex. More and more particles were discovered, all of which could with equal justification be called elementary. Moreover, the theoretical description of these particles using the older methods no longer appeared to be meaningful. This led to the era of "nuclear democracy," which posited the existence of many particles of equal status, with the existence of any one of them including the existence of the others. However, this theoretical framework was soon seen to be insufficient and possibly inconsistent. Slowly, a theoretical model was developed according to which nuclear constituents such as the proton, the neutron, and many of the newly discovered particles of the nuclear democracy were taken to be composed of simpler constituents, *quarks*. This model was soon developed into a consistent theory, called *quantum chromodynamics*. The development of this "particle zoo" is displayed graphically in Figure 6.15 at the end of the book.

Forty years later, the question about which particles are truly elementary seems easier to answer, and at the same time, the goals are more ambitious. Today, one wants not only to explain the properties of matter starting from the basic properties of elementary particles, but also to deduce the properties of elementary particles themselves from elementary principles. This is the subject of *Dreams of a Final Theory: The Scientist's Search for the Ultimate Laws of Nature*, written in 1992 by the Nobel Prize winner Steven Weinberg. It seems, however, that we now are further away from discovering the "final theory" than we were when Weinberg had his dream.

In the first half of the twentieth century, theory played the leading role, but in the second half of that century, progress was largely initiated by new experimental techniques. The basis for these techniques was also established in the first half of the century, but the enormous development since that time can easily be seen from the increase in the size of research instruments, illustrated in Figure 1.2: The "large" accelerator built by E. O. Lawrence in 1939 had a diameter of 37 inches. The Large Hadron Collider (LHC), under construction at the European Organization for Nuclear Research (CERN) in Geneva, Switzerland, with its diameter of more than five miles, exceeds the limits of the canton of Geneva. There are plans—still to be financed—to construct a linear accelerator (TESLA) with a length of more than 20 miles.

Figure 1.2. Magnet for 37-inch cyclotron (top) and Large Hadron Collider (bottom) depicts how accelerators have increased in size. The top photograph shows Lawrence, on the right, and a collaborator, in the yoke of the magnet for the 37-inch cyclotron. The magnet was obtained at a bargain price from the Federal Telegraph Company because its application in radio transmission had been made obsolete by the development of vacuum tubes. The bottom photograph shows the ring of the Large Hadron Collider (LHC) at CERN in Geneva.

Detectors that were originally the size of cigar boxes, are today as big as houses. The quantity of data flowing from a typical measurement is impressive even to communications specialists. It is no wonder that the Internet was developed at CERN. As a result of such growing complexity, ever larger numbers of scientists are involved in a single experiment. In 1933, C. D. Anderson proved the existence of antimatter. His article in *Physical Review Letters* was four pages long. By contrast, the discovery of the top quark in 1995 resulted from research undertaken by two large groups of scientists. When this discovery was described in print, the list of authors and institutions alone filled nearly four pages.

Modern particle physics began around 1950. Since that time, accelerators have played a dominant role in its development. Now, theory is once again ahead of experimental physics; there are many serious theoretical speculations that that have yet to be verified.

Today, with the standard model of particle physics, we have the means to explain the essential features of the dynamics of elementary particles. Many physicists, in fact, are more worried about the theory being too good, leaving them with no new physics to discover, than about potential problems with the standard model. It is generally agreed that further essential progress can be made only if gravity can be incorporated into the quantum field theory of elementary particle physics. There are fascinating links between the physics of the small and of the large: particle physics and cosmology. Knowledge gained from the study of particle physics is essential for explaining the early history of the universe, and some hypotheses in particle physics can at the moment only be tested by examining their cosmological consequences.

We will meet two constants of nature again and again: the speed of light in a vacuum, denoted generally by the letter c, and the elementary quantum of action, Planck's constant h. The speed of light was first measured in 1676 by the Danish astronomer Olaf Römer, but its overwhelming importance was not realized until much later, when Einstein incorporated it into his special theory of relativity. The speed of light has the same value for all observers, independent of the observer's velocity with respect to the source of the light. No particle can move faster than light, and in particular, no signal can be transmitted faster than light. One of the consequences of this is that time is relative. The same process can have different durations for different observers. This sounds incredible, but it has been well tested empirically. The speed of light is very large compared to our everyday experience: 186,000 miles per second (300,000 km/s). Though Einstein's theory of relativity is valid in general, deviations from our intuitive expectations and experience only become apparent if we consider systems whose speeds are comparable to that of light. Then the laws of Newtonian physics have to be replaced by those of special relativity.

In particle physics, the speed of light in a vacuum is the natural unit for measuring velocities. In this context, "slow" describes moving at a velocity that is much less than the speed of light, while "fast" describes moving at a velocity comparable to that of light.

Planck's constant was introduced in 1900 in order to describe the phenomenon of heat radiation. It is an essential constant for all the processes that occur at the atomic scale and below. It is the fundamental quantum of action, where "action" has special meaning in physics as the product of energy and time. The numerical value of Planck's constant is very tiny when expressed in units of everyday life: it is approximately $6.626 \times 10^{-34} = 0.000\ldots0006626$ J·sec (joule seconds), where the notation indicates that there are 33 zeros after the decimal point. It is highly probable that quantum physics is also valid for macroscopic processes, but if the action is large compared to Planck's constant, one can approximate the constant by zero, and one is returned to classical physics. Angular momentum has the same units as action, and therefore Planck's constant is the natural unit for expressing angular momentum on the quantum scale. For practical reasons, one usually uses the *reduced Planck constant* $\hbar = h/(2\pi) \approx 1.05 \times 10^{-34}$ J·sec.

A unit of energy in particle physics is usually expressed as an electron volt, which is the energy gained by an electron when it passes through an electrostatic potential difference of one volt. For practical reasons, electron volts usually are stated using the measure of mega electron volts and giga electron volts. The mega electron volt (MeV) is equal to one million electron volts, and the giga electron volt (GeV) is equal to one billion electron volts. GeV is expressed as BeV in the older literature.

The customary unit of mass is the mega electron volt divided by the square of the speed of light in a vacuum, MeV/c^2, or the giga electron volt divided by the speed of light, in a vacuum GeV/c^2; more details are given in Appendix B. The appendix also includes a glossary of the most important and widely used terms in this book together with references to the sections in which they are introduced.

1.2   Brave Old World

I begin by describing the history of the discovery of the elementary particles that occur naturally and by explaining the structure of matter under normal circumstances.

The proton. It is impossible to determine exactly when the proton, one of the most important building blocks of matter, was discovered. Today, we know that a single proton forms the nucleus of hydrogen, the lightest chemical element. However, when hydrogen was discovered by Henry Cavendish in 1766, it was by no means clear what a chemical element was, much less what constituted an atom. Most remarkable, however, is the hypothesis that was advanced by William Prout in 1815. Based on the atomic weights that were known at the time, Prout hypothesized that all elements are formed from hydrogen. This hypothesis was very fruitful, even if it turned out later not to be fully tenable.

The electric charge of the proton is sometimes called an *elementary charge*. If one speaks of a particle with an (elementary) charge of -1, one means that it has a charge that is opposite to that of a proton. The mass of the proton is 1.672×10^{-27} kilogram, or in the standard units of particle physics, 938.2720 MeV$/c^2$. Reference to other important properties of the proton will be made later on, in connection with the other building block of the nucleus, the neutron.

The electron. The electron is the first elementary particle about which one can speak of a clear discovery, and it has maintained its status as an elementary particle to this day.

The discovery of the electron was made possible by advances in vacuum technology. The mechanic and glassblower H. Geissler constructed air pumps and discharge tubes that were so well sealed that the air pressure in the tubes was stable at about one ten-thousandth of atmospheric pressure. With these tubes, Julius Plücker, best known as a mathematician, made a thorough investigation of electric discharges. He found that from the negative pole, the cathode, radiation was emitted that resulted in a green glow at the opposite wall of the tube. This effect is still used today, for example, on radar screens and—in a more refined form—in color television.

Heinrich Hertz, the discoverer of radio waves, found that this radiation could penetrate thin aluminum foils, and that led his student, Philipp Lenard, to extract the radiation through such a foil making it easy to experiment with these *cathode rays*. The nature of the rays was obscure. Most British physicists believed them to be beams of particles, while in Germany the idea that they related to the ether was more popular.

If one accepted the concept of a particle beam, one could determine the ratio of the electric charge to the mass by observing the beam's deflection in a combined electric and magnetic field. In January 1897, Emil Wiechert in Königsberg estimated this ratio of charge to mass on the basis of his experiments to be 200 to 4,000 times that of the hydrogen ion. Assuming the same value of the charge, he concluded that there existed a particle that was 200 to 4,000 times lighter than the hydrogen atom.

That same year, W. Kaufmann, in Berlin, and J. J. Thomson, in Cambridge, England, also measured this ratio. Kaufmann obtained the value 1,000, while Thomson obtained the value 770. However, the two researchers reached completely different conclusions. Kaufmann argued that his result was probably irreconcilable with a particle interpretation, since there were no known particles with such a small mass. Thomson, however, concluded that the beam consisted of particles that have either a much smaller mass or a much greater charge (or both) than the hydrogen ion. The final breakthrough came in 1899 when Thomson employed a newly developed instrument, the cloud chamber, which was invented by his student, C. T. R. Wilson. (We will explore the cloud chamber in more detail in Section 1.3.) By counting the droplets in the chamber, Thomson could estimate the electric charge of the particles in the cathode rays, and he obtained a value for the charge that allowed him to conclude that the particles were at least 60 times lighter than hydrogen ions. In fact, the correct number is 1,836 instead of 60. Although Thomson greatly overestimated the mass of the particles, he firmly established the existence of a particle much lighter than the simplest atom—the electron had been discovered. The electric charge and mass of the electron are now known with great accuracy: the charge is exactly one negative elementary charge, and the mass is $0.51099899 \, \text{MeV}/c^2$.

At the time of the discovery of the electron, attention was focused on two properties—charge and mass. With the development of quantum mechanics in the 1920s, another property turned out to be essential—the *spin*. This property is only partially accessible to our intuition, which is, of course, formed by our experience from everyday life. Nonetheless, spin has many properties in common with classical angular momentum, and the formal treatment in quantum mechanics is the same. In some respects, one can consider spin as the angular momentum caused by a classical rotation of the particle. The classical model of a spinning particle can indeed explain some properties, but in other respects this paradigm falls short. The numerical value of the electron spin is exactly half of the reduced Planck constant, $\frac{1}{2}\hbar$. We will return to the topic of spin in more detail in Section 1.5.2.

The electron acts like a small magnet; its dipole moment is—to great accuracy—the value of the spin times the ratio of charge to mass. If the spin were a normal angular momentum, one would expect only half that value.

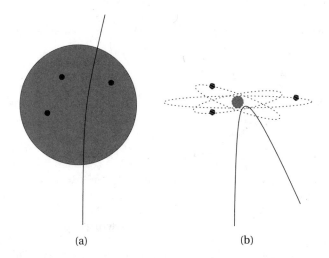

(a) (b)

Figure 1.3. (a) Thomson's model of the atom. In Thomson's model, the electrons (black) swim in the extended, positively charged atom (gray) like "plums in a pudding." (b) Rutherford's model of the atom. In the model of Rutherford and Bohr, the electrons revolve around a small nucleus like the planets around the sun. The solid line is the trajectory of a particle hitting the atom. (The figure is not to scale.)

Models of the atom. After the discovery of the electron, J. J. Thomson developed a model of the atom in which the electrons are distributed like plums in a pudding; see Figure 1.3(a). This model explained some important properties of matter quite well, such as the optical index of refraction. The model became untenable, however, after Ernest Rutherford's interpretation of the experiments of Hans Geiger and Ernest Marsden disproved the model.

Although these experiments did not lead to the detection of new elementary particles, the experiments are important in the history of particle physics. Rutherford's interpretation led to a new branch of physics—nuclear physics, which is the precursor of modern particle physics. In addition, Rutherford's technique of particle scattering was essential in the investigation of the structure of matter, from elementary particle physics to the physics of condensed matter.

Rutherford had investigated the scattering of alpha particles on thin gold foil. An *alpha particle* (α-particle) is the nucleus of a helium atom emitted from a radioactive nucleus. Geiger, who worked at Rutherford's institute in Manchester, continued these experiments and confirmed that alpha particles are only weakly deflected by the gold atoms. This was to be expected, since the fields in

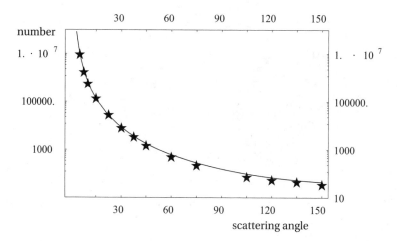

Figure 1.4. Result of the scattering experiment of Geiger and Marsden. The stars indicate the experimental results; the errors are of the same size as the symbols. The solid line is the prediction for a point-like nucleus.

an extended Thomson atom are too weak to deflect the alpha particles strongly. Figure 1.3(a) shows a typical trajectory of an alpha particle passing through an atom. Although Rutherford's and Geiger's experimental results were as they expected, they asked their student Marsden to look for possible events in which the alpha particles were scattered backward, hoping that the further experiments would be informative.

The first results of these experiments were as incredible "as if you fired a 15-inch shell at a piece of paper tissue and it came back and hit you," as Rutherford later remarked. He found the solution: if the positive charge of the atom is not distributed uniformly over the whole atomic volume, of size about one cubic nanometer, but concentrated in a much smaller nucleus, then the electric field very near that nucleus will be strong enough to reflect an alpha particle, even if its energy is around 8 MeV, a very high energy for the standards of the time. This is illustrated in Figure 1.3(b). The angle made by the particle as it is deflected from its path is called the *scattering angle*. The scattering angle in Figure 1.3(a) is small, while that in Figure 1.3(b) is nearly 180 degrees.

Geiger and Marsden performed the tedious experiment very thoroughly, and their results are summarized in Figure 1.4. They found that most alpha particles are only weakly deflected, but a few are scattered backward, that is, with a scattering angle larger than 90 degrees. The solid line shows the theoretical curve for the case in which the whole charge is concentrated in a nucleus whose entire mass is concentrated at a point. The small deviations of the measured points from the curve indicate that the nucleus has finite extension. From the data, one could conclude that the diameter of the gold nucleus is less than one-thousandth the diameter of the atom; today we know that it is about 30,000 times smaller.

I will digress now to explain something of great importance for scattering theory: the *cross section*. Let us assume that there are n scattering centers that do not influence one another; these form the "target." Let us also assume that there is an incident particle beam with a flux of j particles per unit of time and area. The cross section of a scattering center—which in the experiment of Geiger and Marsden was the nucleus of a gold atom—is defined as the number of scattered particles divided by the product of the particle flux times the number of scattering centers, that is, $N/(jn)$. The curve shown in Figure 1.4 shows how the cross section of the gold nucleus changes as a function of the scattering angle.

In 1913, Rutherford's atomic model motivated Niels Bohr to propose an alternative atomic model in which the electrons orbit about the nucleus like planets around the sun. However, according to the well-established laws of classical physics, Bohr's model was not tenable, since an orbiting electron emits radiation and thereby loses energy. Sooner or later the electron would collapse into the nucleus. Bohr concluded that classical electrodynamics fails at the atomic level and that for special, *quantized* orbits, the electrons do not emit radiation.

According to Bohr's model, electrons can emit or absorb energy only if they jump from one quantized orbit to another. Such a transition is called a quantum jump and represents the smallest possible change in energy. When an electron jumps from one orbit to another, it emits a quantum of light, a *photon*, whose energy is the energy difference of the two orbits. According to Einstein, the frequency of the emitted radiation is the energy of the photon divided by Planck's constant.

Louis de Broglie, Werner Heisenberg, Wolfgang Pauli, Erwin Schrödinger, Max Born, Paul A. M. . Dirac and others later set up the first formalism of the "new" quantum mechanics in which Bohr's formulas could be derived consistently. According to the formalism of quantum mechanics, the possible states have the same energy as Bohr's allowed quantized orbits. Quantum mechanics made many more predictions in all the subfields of physics. These predictions sometimes completely contradict our expectations, but all have been tested experimentally.

More recently, many of the most paradoxical results of quantum mechanics have been shown to be correct. Wherever the hypotheses of quantum physics and our intuitive beliefs diverged, experiments have always confirmed the truth of the scientific hypotheses. Despite the fact that quantum mechanics does not confirm the intuitive picture of orbiting electrons in the atom, quantum mechanics does explain why the simple Bohr model works in some special cases.

Essential for understanding atomic physics and especially the periodic table of chemical elements is Wolfgang Pauli's exclusion principle: if a quantum-mechanical state is occupied by an electron, no other electron can occupy the same state. The exclusion principle applies only to particles with half-integer spin. These particles are called *fermions*, for reasons that will become clearer in Section 2.4.,

The neutron. In 1930, Walther Bothe and H. Becker discovered a new hard (penetrating) radiation, that occurs when the element beryllium is bombarded by alpha particles. It was so hard that there was little doubt that it originated in the atomic nucleus. The couple Irene Joliot-Curie (daughter of Pierre and Marie Curie) and Frédéric Joliot investigated this radiation, confirming many of the results of Bothe and Becker. However, Joliot-Curie and Joliot also discovered some essential new characteristics of the radiation, which seemed to defy a simple explanation. For them, too, the nature of the radiation remained a mystery.

Earlier, in 1920, Rutherford had speculated about the existence of a very tight, electrically neutral, bound state of a proton and an electron. He assumed that this state should have the approximate size and mass of a proton. He came to these assumptions through his study of astrophysics. Rutherford could not understand how elements could be formed inside stars without the help of heavy neutral particles. Meanwhile, James Chadwick, a close collaborator of Rutherford, was looking for these neutral particles. Inspired by the investigations of Joliot-Curie and Joliot, Chadwick soon discovered that the radiation identified by Bothe and Becker consisted precisely of these neutral particles; he had discovered the *neutron.*

Before the discovery of the neutron, it was assumed that the atomic nucleus consisted of protons and electrons. This assumption explained why the mass of an atom is approximately an integer multiple of the mass of a hydrogen atom. (Recall that the mass of the electron is negligible compared to that of the proton.) This hypothesis also explained radioactive beta decay, a process in which an atom emits light, negatively charged particles. That these particles are indeed electrons had been shown by Kaufmann in 1901.

However, the notion that the nucleus is composed of electrons and protons led to theoretical difficulties. According to quantum mechanics, it was impossible to confine light objects such as electrons within the small volume of the atomic nucleus. Immediately after the neutron was discovered, Heisenberg proposed that the nucleus is composed of neutrons and protons.

Like the electron, the proton and the neutron have spin $\frac{1}{2}$ and—as was found later—a magnetic moment. The magnetic properties of the proton, and of nuclei in general, play an important role in medical diagnostics, since they are the basis of nuclear magnetic resonance imaging (MRI).

As an isolated particle, the neutron decays radioactively into a proton, an electron, and one additional particle, which will be discussed later. When the neutron is bound in an atomic nucleus, it can be stable. The neutron has approximately the same mass as the proton; it is heavier by roughly one-tenth of one percent. This fact led Heisenberg, in 1932, to conjecture that these two particles are connected by a symmetry, which later was called *isospin*. Isospin will be discussed in Section 1.5.3.

Neutrons occur in large concentration in neutron stars. Due to their rotation and their high magnetic moment, neutron stars emit electromagnetic signals analogous to the revolving lamp in a lighthouse. From the very short period of 30–100 milliseconds that the signals are emitted, one can conclude that the diameter of such stars is very small, only around 20 km.

Though Rutherford conjectured that neutrons exist, and though this assumption was very important for Chadwick's search for a neutral particle, one cannot say that the neutron was theoretically predicted. This was not the case for the photon and the neutrino, two particles discussed below.

The photon. The question of the nature of light had puzzled scientists since the beginning of modern scientific inquiry and led to heated debates. Adherents of the corpuscular theory, such as Newton, assumed that light consists of small particles. Others, adherents of the wave theory such as Chistiaan Huygens and Leonhard Euler, considered light to be a wave phenomenon. In the beginning of the nineteenth century, the wave theory won out, gaining wide acceptance as a result of the work of Thomas Young and Augustin-Jean Fresnel, among others.

Indeed, James Clerk Maxwell, through his unified theory of electricity and magnetism, was able to explain light as an electromagnetic wave phenomenon and to deduce all the essential properties of light. In 1887, Hertz gave even more support to the wave theory by providing unequivocal proof of the existence of electromagnetic waves. The nature of light seemed finally to be settled: it was due to oscillations within some medium, which was called the *ether*.

But in 1905, a twenty-six-year-old technical expert at the patent office in Bern named Albert Einstein published three papers that fundamentally changed the course of physics in the twentieth century. One of Einstein's papers was titled: "On a Heuristic Viewpoint Concerning the Production and Transformation of Light." In this paper, Einstein summarized his findings by stating: "Following the assumption considered here, the energy of a light beam going out from one point is not continuously distributed over larger and larger spaces, but the energy consists of a finite number of energy quanta localized in space points. These move without splitting themselves and can only be integrally absorbed or created." Today, these quanta are called *light quanta*, *gamma quanta*, or *photons*. The mass of the photon is zero, and its energy is proportional to the frequency of the associated light waves; the proportionality constant is Planck's constant.

With his heuristic point of view, Einstein could explain the mysterious property known as the *photoelectric effect*. Heinrich Hertz already had discovered that when a beam of light is directed at certain metals, electrons are ejected from the metal. Lenard, a student of Hertz, had investigated this effect further and found that the energy of the ejected electrons does not depend on the intensity of the incident light, that is, on the flow of energy, but on its wavelength. The smaller the wavelength and thus the higher the frequency, the greater the energy of the electrons. If the frequency is lower than a certain threshold value, then no electrons are ejected at all.

This mysterious feature is easily explained by Einstein's hypothesis of light quanta: an atom of the metal absorbs a light quantum (photon). Part of the energy of the photon liberates an electron from its atomic bond, and the rest of the photon's energy gives the electron its kinetic energy. According to Einstein, the photon's energy is proportional to the frequency; if it is too small, the energy is insufficient to overcome the binding energy and eject an electron from the atom. This explains why ultraviolet light, with its relatively high frequency, is chemically much more aggressive than visible light.

Einstein was able to explain the results of photoelectric emission not only qualitatively, but also quantitatively. The energy of the photon is given by the product of the frequency f and Planck's constant h. The energy $E_{electron}$ of the ejected electron is therefore given by

$$E_{electron} = hf - B,$$

where B is the binding energy of the electron in the atom. Later, Einstein derived Planck's radiation formula elegantly in the context of light quanta.

In spite of its success in explaining the mysterious features of the photoelectric effect, the photon hypothesis at first found very little acceptance, and the cautiously formulated title of his paper shows that Einstein was well aware that such would be the case. The continuous wave nature of light seemed to have been established beyond doubt. In the 1920s, Arthur Holly Compton performed scattering experiments with x-rays, and the results could be interpreted only by assuming collisions of corpuscular photons with electrons. When finally, through the further development of quantum mechanics by de Broglie, a wave nature also was attributed to well-established particles such as electrons, the concept of the photon became generally accepted.

Thus, when we speak throughout this book of "particles," we are referring to objects that are described by the language of quantum physics; we will return to this topic in Section 1.4.2 and in the conclusion. It is merely a question of definition as to whether one wishes to view photons as part of matter, but undoubtedly they are an essential part of our universe. Photons form by far the largest number of (known) particles in our universe.

The neutrino. Wolfgang Pauli called the neutrino the "foolish child of my life crisis that continued to behave foolishly." Even today, the neutrino continues to defy researchers' assumptions. In 1930, Pauli postulated the existence of a new neutral particle for two reasons. The first reason is that in radioactive beta decay, the nucleus emits an electron, whose spin, it had been found, changes by one unit, whereas the electron has spin of only $\frac{1}{2}$. The second reason, which is even more compelling, is that the emitted electrons do not have a fixed energy, but rather a continuous energy distribution, stretching from zero to a maximal energy typical of the specific type of decay. If the mother nucleus decayed into only two particles—the daughter nucleus and one electron—then the electron would have a fixed energy, because of the conservation of energy and momentum. This is indeed the case in alpha decay, in which the mother nucleus emits only one particle.

Pauli postulated that in beta decay, not only the electron is emitted, but also a further light, neutral particle, which he initially called *neutron*. Later, after Chadwick's discovery of the heavy neutral particle, Pauli called the light neutral particle the *neutrino*. The neutrino should be much lighter than the electron, perhaps even without mass, like the photon, and should have spin $\frac{1}{2}\hbar$. This hypothesis simultaneously solved the two problems mentioned above: the neutrino and electron together carry an integer spin, and they can share the energy liberated in the decay; consequently, the continuous energy distribution can be explained.

The neutrino was not discovered until 1956; this topic will be explored in Section 2.8. Further on, in Section 7.1, we will see that neutrinos have again become a focus of scientific interest.

1.3   Detection of Particles

Modern particle physics differs in one fundamental respect from earlier approaches. In modern particle physics, one does not merely speculate about elementary particles; one tries to detect them. Therefore, particle detectors play an essential role, and progress in particle physics has been made possible to a large extent by the development of increasingly efficient detection devices.

In most cases, ionization is the basis for the detection of a particle. A charged particle, e.g., a proton, knocks an electron out of an atom. It splits the neutral atom into a free electron and a positively charged remainder, called an ion. This process has many consequences:

1. It can lead to chemical reactions, such as those that take place in the emulsion of a photographic plate.

2. If the ions recombine with electrons, light can be emitted.

3. Gases and liquids become conductors through the separation of the charges.

4. The ions form condensation seeds for droplets or bubbles in gases or liquids.

The ionization processes cause charged particles to lose virtually all their energy as they travel along their path in the detector. This allows a determination of the particle's energy. The loss of energy per unit of length can also be measured. Since the loss of energy depends primarily on the velocity of the ionized particle, the velocity can be determined by counting the ion density.

Photons are electrically neutral, but they can nevertheless ionize atoms by the photoelectric effect discussed above. Therefore, they can be detected in the same way that charged particles are detected. Another important process for detecting photons is pair creation: a photon is transformed into an electron and a *positron*, a positively charged particle with the mass of an electron. This interesting process can be explained only in the framework of quantum field theory, as we will see in the next section.

Recently, the *Cherenkov effect* has been used for detecting charged particles. This effect is very similar to the emission of a shock wave by a body, such as an aircraft, whose velocity in air exceeds the speed of sound. In the Cherenkov effect, a light wave is emitted; in order to emit Cherenkov radiation, the velocity of the radiating particle must be greater than the velocity of light in the medium through which the particle is moving. Pavel Alekseyevich Cherenkov, Ilya Mikhailovich Frank, and Igor Evgenyevich Tamm were awarded the 1958 Nobel Prize in physics for the discovery and theoretical explanation of Cherenkov radiation.

I will now describe some detectors that have played an important role in the development of particle physics.

Photoemulsion. The photographic plate is the oldest particle detector. In 1896, Henri Becquerel discovered radioactive rays after they blackened a photographic plate wrapped in black paper. The particles of the radioactive rays trigger, through ionization, the same chemical reactions as those caused by light. If the exposed photographic plate is developed, the trajectory of an ionizing particle becomes visible as a black track in the emulsion. This method was further developed in Vienna and culminated in the production of a special *nuclear emulsion* by the firms Illford and Kodak. We owe important discoveries in particle physics to this old method.

Fluorescence and scintillation counters. Some substances emit visible light if the ion recombines with an electron. This process is called *fluorescence.* The electrons of cathode rays in Geissler's tubes became noticeable as green light, which is the fluorescence light in the glass. Wilhelm Conrad Röntgen discovered x-rays when they caused a platinum compound to fluoresce.

The scintillation counter is also based on fluorescence, detecting the flash of light produced by a single particle. In the pioneering days, the counting was done by eye. In the classic experiments of Geiger and Marsden, described in Section 1.2, the scattered alpha particles were painstakingly detected in a visual scintillation counter. In modern counters of this type, the emitted light is measured by photovoltaic cells and registered electronically.

Ionization chamber. Soon after the discovery of radioactivity, Becquerel found that radioactive radiation makes air conductive and leads to the discharge of an electroscope. In the ionization chamber, the radiation ionizes the contained gases, and one can measure the current that is conducted by the

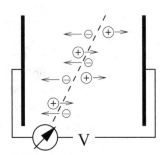

Figure 1.5. The atoms and molecules along the trajectory (dashed line) are ionized by a charged particle. The applied voltage V attracts the positive ions to the negative pole, the electrons to the positive one. The current is registered, and from the time behavior of the current, one obtains information about the ionizing particle.

separated charges; this is represented in Figure 1.5. Earlier, the current was measured through the discharge of an electroscope; today, it is amplified and registered electronically.

The ionization chamber has played an important role since the early days of particle physics. Chadwick discovered neutrons with the help of an ionization chamber. It was lined with paraffin, and the hydrogen nuclei in the paraffin were bombarded by neutrons. Protons that were struck and accelerated by the neutrons then were detected by ionization effects.

The Geiger counter. The best-known particle detector is the Geiger counter, also called the Geiger–Müller counter. It consists of an ionization chamber in which the electric field in certain regions is very strong. In a cylindrical tube with a metal column in the middle, the field strength increases strongly toward the center. The electrons from the primary ionization are accelerated by the electric field. If they are fast enough, they can ionize other atoms or molecules and the number increases by a chain reaction; see Figure 1.6.

The surge of current released by the ionizing particle can easily be registered. One can amplify the current and hear it over a loudspeaker; this leads to the familiar clicking as a telltale sign of radioactivity.

In order to limit the chain reaction and to avoid a permanent current in the counter, a small amount of polyatomic molecules, such as alcohol vapor, is added. According to folklore, this technique was discovered by accident— one of Geiger's collaborators was an alcoholic, and his counters always worked best. If the applied voltage is not too large, the triggered current is proportional to the released energy, and in this case one speaks of a proportional counter.

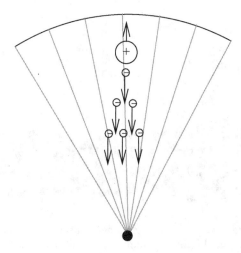

Figure 1.6. Section of a cylindrical Geiger counter. The wall is negatively charged, while the central column is positively charged. The electric field strength increases strongly toward the center. Positive ions are attracted to the walls, as electrons are accelerated in the strong electric field and perform new ionization processes.

A very important method in the application of counters is the coincidence method, which was developed beginning in 1925 by W. Bothe and his collaborators. Bothe and Geiger investigated Compton scattering—the scattering of photons on electrons—with two counters. They found that the scattered photon and the electron simultaneously triggered the two separated counters. In the first experiments, the two electroscopes of the counters were filmed. Later, Bothe developed an electronic coincidence circuit that was improved greatly by Bruno Rossi in 1930. Coincidence and anticoincidence methods continue to play an essential role in particle detection and identification.

The cloud chamber. The development of the cloud chamber by C. T. R. Wilson can be traced directly back to the discovery of the electron by J. J. Thomson. Wilson had been investigating the influence of electric charge on vapor pressure. Above a flat surface of water in a closed receptacle at fixed temperature, the molecules in the liquid and in the vapor are in dynamic equilibrium, determined by two concurrent processes. In one process, the thermal motion of the molecules causes some molecules to escape from the liquid. In the other process, the attractive force of the molecules in the liquid prevents the molecules from escaping too often and attracts some molecules from the vapor.

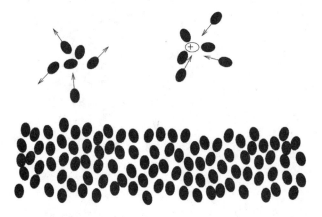

Figure 1.7. Principle of Wilson's cloud chamber. In saturated vapor, as many water molecules enter the liquid as leave it (bottom). Even in supersaturated vapor small droplets cannot form, since there are not enough neighbors to hold the droplet together (top left). An ion serves as a condensation seed. Since it attracts water molecules, a droplet can form (top right).

On a strongly curved surface, a molecule has fewer neighbors than on a flat surface and therefore is retained less strongly in the liquid. Thus, a small droplet evaporates when a flat surface is in equilibrium with the vapor. The smaller the droplets, the fewer neighbors their molecules have; therefore evaporation is strongest for small droplets.

One can produce supersaturated water vapor in which there are more water molecules than there should be according to the equilibrium. If the gas is absolutely pure, no small droplets can form, since they would evaporate immediately. Since all droplets have to start small, no droplets are formed at all. However, the electric charge of an ion attracts water molecules, which can lead in a supersaturated vapor to the formation of droplets, since the attractive electric force prohibits immediate evaporation. Therefore, ions form seeds for the condensation of water droplets.

This process had been investigated theoretically by Thomson, and he derived the relevant formulas. He could even derive the charge of the electron from the specific droplet formation, as mentioned in Section 1.2. From these principles, Thomson's student C. T. R. Wilson developed one of the most important detection devices in particle physics, the cloud chamber.

The principle of the cloud chamber's construction is simple. The temperature in a receptacle with saturated water vapor is lowered by expanding the container's volume. Droplets can form only at ions, and since a charged particle creates ions along its trajectory, this trajectory becomes visible through the water droplets forming at the ions. By a suitable choice of operating conditions, the density of the tracks makes it possible to determine the velocity of the particle. The track is photographed with two cameras stereoscopically so that it can be reconstructed in space.

A significant advance in the technique was made by P. M. S. Blackett and G. P. S. Occhialini. They surrounded the cloud chamber with counters and expanded it only if the counters indicated that at least one charged particle had traversed the chamber. In this way, a picture was taken only if there was something—perhaps something interesting—to be seen. Though this seems to be a rather simple concept, it required one to possess an instinctual certainty in order to put the concept into operation. Blackett, the cloud chamber specialist, disparaged the Geiger counter: "In order to make it work you had to spit on the wire on some Friday evening in Lent."

Many important discoveries in early particle physics were made with the cloud chamber. It also played a role in early discussions about quantum mechanics. Critics of Heisenberg raised the following question: if in quantum mechanics the uncertainty principle prohibits a particle from forming a classical trajectory, why does it leave a visible trace in the cloud chamber? Heisenberg's answer was that the trace in the cloud chamber is not a trajectory in the mathematical sense, and the size of the droplets indicates some uncertainty about the position of the ionizing particle.

1.4 Quantum Physics Becomes Decisive

Quantum physics was developed in order to describe atomic phenomena. For a time, it was unclear whether it would also be applicable to subatomic phenomena, for example, to processes in the atomic nucleus. It was widely believed that, for nuclear physics, new laws had to be found, which might be as different from those of quantum physics as the laws of quantum physics are from those of classical physics. It became clear that quantum physics is generally valid. However, it is not quantum mechanics but quantum field theory that must be applied in particle physics. Quantum field theory is the quantized form of classical field theory. We will discuss quantum electrodynamics, the quantized version of classical electrodynamics, later.

In this section, which is not organized in strict historical order, I will try to describe some of the essential results of quantum field theory, using as little formalism as possible. This leads necessarily to a balancing act between oversimplification and incomprehensibility, but I think that some essential ideas can be successfully conveyed. The few formulas that I present might seem complicated because of the unfamiliar symbols they contain, but I will not make use of higher mathematics. I will try to speak only the truth, but I cannot always tell the whole truth.

1.4.1   Special Relativity and Quantum Physics

We begin with one of the most important relations discovered in the last century: the general relation between the energy E, momentum $\vec{p}$, and mass m of a particle and the speed of light c:

$$E^2 = m^2 c^4 + \vec{p}^2 c^2.$$

This relation was established by Einstein in 1905.

This formula was doubly revolutionary. First, for particles at rest, that is, with negligible momentum, it yields the famous relation for the rest energy:

$$E = mc^2.$$

Second, it has two solutions for a given mass and momentum, a positive and a negative one:

$$E = +\sqrt{m^2 c^4 + \vec{p}^2 c^2} \quad \text{and} \quad E = -\sqrt{m^2 c^4 + \vec{p}^2 c^2}.$$

In classical physics, the negative solution is of no interest; one declares it unphysical and ignores it. But in quantum mechanics, matters are not so simple. In quantum mechanics, one assigns certain *operators* to observable properties such as position, momentum, and energy. This differs from classical physics, in which functions are assigned to these observables, the values of which are the specific results of measurement. Operators are identified by their actions, not by their values. The energy of a state is determined by the way in which the energy operator acts on the state. If certain mathematically possible results of a measurement, such as the negative energy values mentioned above, are disposed of, this action modifies the operator. It turns out that this modification has consequences that contradict the principles of special relativity. One of the principles that is violated if the negative energies are left out is *locality*. Since this principle plays an important role in particle physics, I would like to discuss it here briefly.

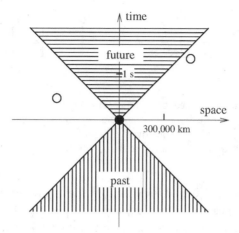

Figure 1.8. Graphical representation of locality. A special event happening at time 0 and at position 0 is marked by a solid circle; the space-time region that can be influenced by this event is horizontally striped; the region that can influence the special event is vertically striped. Events that occur at the time and position indicated by open circles are completely independent of the special event, since they cannot communicate by signals travelling maximally at the speed of light.

The principle of locality means—expressed somewhat loosely—that an event cannot be influenced by another event of which it can have no knowledge. In particular, an event cannot be influenced by a future event. Events also are completely independent if they cannot communicate with signals propagating maximally at the speed of light. This is based on the postulate of special relativity that states that signals cannot propagate faster than light.

In Figure 1.8, the space-time domain that the event marked by a solid circle can influence is indicated with horizontal stripes, while the domain that can influence the event is vertically striped. The more distant an event is from the marked one, that is, the farther to the right or to the left it is in the figure, the longer it takes until it can be influenced. An eruption on the sun can influence the earth only after eight minutes, since light needs this time to travel from the sun to the earth. It should be noted that in all reports on teleportation in quantum mechanics with velocities faster than light, one is not dealing with signals that transmit information.

One could tolerate the violation of the locality principle in quantum mechanics, but in the course of the history of science it has turned out that it is generally worthwhile to be conservative and not to surrender physical princi-

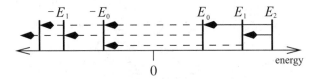

Figure 1.9. The positive and negative energy states of the Dirac equation. In addition to the well-established transitions between states of positive energy (solid arrows) one expects transitions of states with positive energy into those with negative energy and also transitions between states of negative energy (dashed arrows).

ples too easily. It is frequently the case that sticking to principles has led to revolutionary changes in other respects, which is just what occurred here.

In 1928, P. A. M. Dirac had established a quantum-mechanical equation for the electron that respects the relation between energy, mass, and momentum that is discussed above. This equation simultaneously solved many puzzles in atomic physics. It showed that electrons must have spin $\frac{1}{2}\hbar$ and that the ratio of magnetic moment to spin is larger by a factor of two than expected by classical physics, Dirac's equation also explained with high precision the observed lines in the hydrogen spectrum, that is, the wavelengths of light emitted by hydrogen. O. Klein and Y. Nishina used this equation to predict the scattering of photons on electrons. Their results were satisfying as well, though—due to experimental error—not as convincing as in the case of the spectral lines.

However, the Dirac equation posed a tremendous problem that is closely related to the two possible signs of the energy mentioned above. It predicted states with negative energy: for each state with positive energy there existed a corresponding state with an energy of the same magnitude but opposite sign. I. E. Tamm and I. Waller showed independently that these negative states are indeed necessary if one is to obtain (in the limit) Thomson's well-known classical result from the results of Klein and Nishina, that is, by considering photons with long wavelengths. So there was now another reason to take the negative energy states seriously. But if they really existed, then it was unclear why the conventional states, that is, those with positive energy, did not decay into the negative ones and emit light, something that always happens when a state of higher energy is occupied and a state of lower energy is available.

In Figure 1.9 the situation is displayed graphically. The solid arrows signify the observed transitions between states of positive energy that lead to the observed spectral lines emitted by atoms. The dashed arrows indicate the formally possible but not observed transitions in which the final state is one of negative energy. The radiation emitted in the latter case can be considerable. The lowest positive-energy state is that of rest energy, approximately half a million electron volts. In addition, there is, according to the Dirac equation, a state with an energy of negative half a million electron volts. In a transition from the lowest positive to the highest negative state the emitted photon would thus carry an energy of one million electron volts. In order to produce such energetic x-rays, the voltage in an x-ray tube would have to be higher than one million volts. But this is not enough: since there is no state of lowest energy, decay would continue to states of lower and lower energy, and there would be no stable matter at all.

It is no wonder that Dirac's theory was met with sharp criticism. Pauli invented the second Pauli principle: such a theory should be applied to the body of its inventor. He then would decay immediately and could not propagate such a theory. But there was a dilemma: on the one hand, the Dirac equation made extremely precise and well-established predictions, while on the other hand, it led to absurd conclusions such as those just mentioned. In this situation, three alternatives were possible: one could totally reject the Dirac equation, one could assume that the Dirac equation has some meaning but that one has no idea what it is, or one could invent an ingenious explanation that preserved the good features and finessed the negative consequences.

It would have been reasonable to wait for a better interpretation, especially since conventional nonrelativistic quantum mechanics had (and, as some claim, still has) significant problems. But it is not in the nature of physicists to wait when there are such pressing problems to be solved., Fortunately, Dirac himself found an ingenious explanation. The (first) Pauli exclusion principle was already known: if a state is occupied by an electron, there is no room for a second one. Taken together with Bohr's atomic model, this explains the periodic table of the elements. Thus, for example, the electron responsible for the yellow light emitted by sodium does not fall into a state of even lower energy because such states already are occupied.

Dirac now proposed that all the negative energy states of the Dirac equation are present but already are occupied, and therefore a transition to them was impossible. Thus in the modified theory, matter was stable again and behaved as though the negative states were not present. However, Dirac had to assume that all the occupied negative energy states—the so-called Dirac sea—were totally unobservable. Something new, however, could happen: one should be able to observe if one state was *not* occupied; the hole in the Dirac sea of electrons would behave like a particle with opposite, that is, positive, charge. Since until then only two kinds of charged particles were known, the negative electron and the positive proton, Dirac proposed to identify the holes with the protons. He assumed that the interaction with the Dirac sea would give the hole the much higher mass of the proton, which is about 2,000 times that of the electron.

This theory also failed to meet an enthusiastic welcome, and it, too, led to a variety of reactions, from each according to his temperament. Heisenberg reported that "the magnetic electron of Dirac" had driven P. Jordan—one of the most mathematically minded fathers of quantum mechanics—to melancholy. In Rome, Enrico Fermi held a mock trial in which Dirac was (symbolically) condemned for having offended one of the most important principles of quantum physics— the prohibition on the introduction of unobservable quantities. Niels Bohr invented another how-to-catch-wild-animals story, namely, "how to catch elephants." At the elephants' waterhole, a poster is displayed with Dirac's theory. The elephant, a very sage animal, reads the poster and is in total shock for several minutes. During this time the hunter can tie up the elephant and ship him off to the zoo.

When it became clear, in 1931, that the identification of the hole with the proton was untenable, Dirac made a cautious, but clear, statement: "A hole, if there were one, would be a new kind of particle, unknown to experimental physics, having the same mass and opposite charge of the electron." One year later, the experimental physicist C. D. Anderson published an article with the equally cautious title, "The Apparent Existence of Easily Deflectable Positives." By the term "easily deflectable," Anderson meant that the mass was much smaller than that of the proton. Thus the particle corresponding to the hole in the Dirac sea was discovered; we shall return to this topic later. De Broglie proposed that the holes be called *antiparticles*.

1.4.2   Field Theory and Quantum Physics

We come now to the most complex part of this theoretical section, a short description of some essential features of relativistic quantum field theory, a theory that unites the concept of a field with the principles of quantum mechanics.

Generally speaking, a *field* is the assignment of space-time points to certain properties, or as a mathematician would say, a *mapping* of space-time points to these properties. For example, a weather map showing air pressure represents a field: for a fixed time, say August 4, 1913, it displays the air pressure for each point on the earth's surface. Here the air pressure is the physical property: the field strength. In this case we have a material carrier of the field, namely the air. In electrodynamics, in contrast, one became accustomed to considering physical properties such as the electric and magnetic field strength without the presence of a material carrier.

The relationship among the values of a field at different space-time points is expressed through *field equations.* Some well-known field equations are the Maxwell equations of electrodynamics and the Navier–Stokes equations of hydrodynamics. The field equations allow one to calculate future field strengths from current values.

Typical for field theories are superposition phenomena. If a source creates a certain field at a certain point, then this field can be made to cancel the field produced by another source. Such superposition phenomena occur in acoustics, where two tones can produce audible beats, and in optics, where one can observe interference patterns.

In quantum field theory one starts, as in quantum mechanics, from observables such as positional coordinates, momentum, and energy. In classical field theory, as developed by Euler and Lagrange, one can view the field strength as a generalization of the positional coordinates, and express the field energy as a function of the field strength. One can also introduce an analogue to momentum, the *canonically conjugate field momentum* (not to be confused with the momentum carried by the field).

I already have mentioned that in quantum mechanics, operators are assigned to the observables (observable quantities) such as energy and momentum. A special feature of these operators is that they are not *commutative*: if the position operator in space acts on a state, and then the momentum operator acts on the result of this first operation, the outcome will be different from the case in which first the momentum acts and then the space operator.

If we denote the position operator by $\mathbf{X}$ and the momentum operator by $\mathbf{P}$, then one of the fundamental postulates of quantum mechanics is the following *commutation relation*:

$$\mathbf{X} \cdot \mathbf{P} - \mathbf{P} \cdot \mathbf{X} = i\hbar.$$

The product $\mathbf{X} \cdot \mathbf{P}$ means that first the momentum operator $\mathbf{P}$ acts on some state on the right, and then the position operator $\mathbf{X}$ acts on the resulting state. The commutation relation can be simply expressed in the following way: from some state s two new states are formed, one by letting $\mathbf{P}$ act first followed by $\mathbf{X}$, that is, $\mathbf{X} \cdot \mathbf{P} s$; the other one by applying the operators in reverse order, that is,

P·Xs. If one takes the difference of the two results, that is, **X·P**s − **P·X**s, one obtains the original state s multiplied by the number $i\hbar$. This is true for any state, and therefore s is omitted in the commutation relation given above. The number $i\hbar$ is the product of the reduced Planck constant $\hbar$ and the imaginary unit i, that is, a number that multiplied by itself yields -1. The commutation relation of the position and momentum operators leads to Heisenberg's famous *uncertainty principle*. This principle says that it is impossible to measure position and momentum simultaneously with arbitrary precision. The more precisely momentum is measured, the less precise is the outcome of the position measurement, and vice versa.

The quantization procedure of mechanics is now transferred to field theory. In quantized field theory, the fields are no longer directly measurable quantities, that is, operators rather than numbers. The same commutation relations are structurally assigned to these operators and to the canonically conjugate field momentumas as those assigned to position and momentum in quantum mechanics. Wave equations, such as Maxwell's equations, remain unchanged, but the objects occurring in them have changed their meaning—numbers are changed to operators

This is conceptually and technically quite complicated, but I want to emphasize that the principles developed in quantum mechanics together with classical Euler–Lagrange field theory formed an extremely powerful heuristic guide. So one was not, as at the beginning of quantum mechanics, proceeding cautiously in the dark. Therefore, the attempts of O. Klein and P. Jordan to quantize a classical field theory (the second quantization) were crowned by success only two years after the quantization of classical mechanics. However, it took another 20 years before the powerful tool of renormalized perturbation theory was developed for performing quantitative calculations.

The only case in which the quantization of fields has been completely solved is that of *free* fields, that is, fields without any interactions. At first glance, this seems of little interest, but nevertheless, important features of a quantum field theory can be studied for this case. For example, one can achieve a satisfactory solution to the problem of negative energy states and antiparticles. Above all, free-field theory is the starting point for a perturbative treatment of interacting fields. One starts from a free theory and considers the interaction as a small perturbation. Most quantitative results in quantum field theory are obtained by perturbation theory. Later, in Section 6.7, we will investigate a method for going beyond perturbation theory.

Maxwell's theory without sources (charges and currents) is such a free theory: the quantization of this theory has been completely clarified, though it is by no means trivial. Since here we are interested only in general structures, I shall simplify the notation and designate the electromagnetic potential at the space-time point $(\vec{x}, t)$ generically by $A(\vec{x}, t)$. The quantum-mechanical operator that corresponds to the classical electromagnetic potential consists of two parts, which are related by a simple mathematical operation, the Hermitian adjoint:

$$A(\vec{x}, t) = a(\vec{x}, t) + a^*(\vec{x}, t).$$

The Hermitian adjoint of an operator, indicated by an asterisk (*), preserves algebraic relations, and a twofold application leads back to the original operator: $\left(a^*\right)^* = a$. From this, one sees that the field operator $A(\vec{x}, t)$ defined above is *self-adjoint*, that is, its Hermitian adjoint is the operator itself: $A^*(\vec{x}, t) = A(\vec{x}, t)$. The commutation relations that must be satisfied by the operator A according to the above-mentioned quantization rules support the following interpretation of the two parts a and a^*: $a^*(\vec{x}, t)$ is a *creation operator*, that is, it creates a photon at position $\vec{x}$ at time t; $a(\vec{x}, t)$ is an *annihilation operator*, that is, it annihilates a photon at $(\vec{x}, t)$. Thus, the operator a^* applied to a state containing no photons transforms that state into a state containing one photon; the annihilation operator a applied to a state containing three photons transforms that state into a state containing only two photons. It is not astonishing that such operators occur, since, after all, in atomic processes photons are emitted, i.e., created or absorbed, which is to say, annihilated. Since such creation or annihilation cannot be described in quantum mechanics, such fundamental processes as the transition of an excited hydrogen atom into its ground state can be described adequately only in quantum field theory.

The concept of quantization by creation and annihilation operators is more daring for electrons than for photons. It is perhaps no accident that it was consistently performed by E. Fermi in 1934 in establishing the quantum theory of radioactive beta decay. In the nuclear model followed by researchers until 1932, it was assumed that the atomic nucleus consisted of protons and electrons. This led to some difficulties, as described in Section 1.2, but the appearance of electrons in the final state was easily explained—they only had to escape from the nucleus. But when it became evident that the nucleus did not contain electrons but rather protons and neutrons, one had to consider beta decay as a genuine creation process of electrons.

The electron and the positron are described by one and the same field operator, consisting of a creation operator and an annihilation operator:

$$\psi(\vec{x}, t) = b(\vec{x}, t) + d^*(\vec{x}, t),$$

where b annihilates an electron and d^* creates a positron. According to the properties of the Hermitian adjoint mentioned above, the Hermitian adjoint operator $\psi^*(\vec{x}, t)$ is given by

$$\psi^*(\vec{x}, t) = b^*(\vec{x}, t) + d(\vec{x}, t).$$

It contains b^*, which creates an electron, and d, which annihilates a positron.

Since the field operator $\psi(\vec{x}, t)$ has to satisfy the Dirac equation, it must accommodate both positive and negative energy states. This is indeed the case, but the positive energies are assigned to the annihilation operators and the negative energies to the creation operators. If one calculates the field energy, this difference in assignment leads to a positive definite field energy.

In the electromagnetic potential, only one kind of operator occurs, that of the photon, whereas in the electron–positron field ψ there are annihilation operators b for the particle (electron) and creation operators d for the antiparticle (positron). One can summarize this by saying that for the photon, the particle and antiparticle are the same, but such is not the case for the electron. The particles that occur in the quantum fields are called *field quanta*; the photon is thus the field quantum of the quantized electromagnetic field, while the electron and positron are the field quanta of the electron–positron field $\psi(\vec{x}, t)$.

Things become really interesting when we consider interactions. In quantum electrodynamics (QED), that is, the theory of electrons, positrons, and the electromagnetic field, the interaction operator is fixed by general principles, which will be discussed in Section 5.1. It consists of the product of three field operators:

$$L_{\text{interaction}} = e\,\psi^*(\vec{x}, t)\,A(\vec{x}, t)\psi(\vec{x}, t).$$

The strength of the interaction is given by the electron's electric charge e. Moreover, additional factors occur that complicate matters, but are of extreme importance for quantitative calculations, though not for our general considerations here.

I will now skip 20 years of intensive work devoted to efforts to develop these expressions and introduce directly the ingeniously simple graphical method developed by Richard Feynman for writing expressions for quantum-

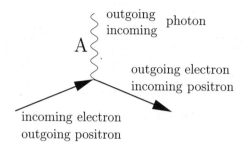

Figure 1.10. Graphical representation of the interaction of photons with electrons (and positrons).

mechanical probability amplitudes. From these, cross sections or decay probabilities can be calculated by squaring the absolute value.

We represent the interaction given above by a vertex graph, as shown in Figure 1.10. The wavy line denotes the photon field A and thus represents creators and annihilators and corresponding photons that move into the vertex and those that pass out of it. The solid line with the arrow pointing toward the vertex denotes the field ψ. It thus represents an incoming (annihilated) electron and an outgoing (created) positron. The line pointing away from the vertex denotes ψ^*. It represents electron creators and positron annihilators and therefore an outgoing electron and an incoming positron.

In order to describe the scattering of a photon on an electron, that is, Compton scattering, we need a graph in which an electron and a photon enter and an electron and a photon exit; this corresponds to the experimental situation. We must construct the experimentally realized situation from the elementary interaction, which means that graphically we must form combinations of the graphs displayed in Figure 1.10. The two simplest possibilities are displayed in Figure 1.11.

In Figure 1.11(a), a photon and an electron are annihilated at point S, and a single electron is formally created. This is incompatible with the conservation of energy and momentum, as one can easily check. But in quantum physics the general uncertainty principle holds between energy and time; the product of time resolution and energy resolution is at least as large as Planck's constant. If a state lives for only a short time, then we need not consider it too precisely with respect to conservation of energy.

That is precisely the case here. The intermediate state lives only a short time; the single electron is created at space time point S and shortly afterward

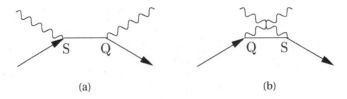

(a)   (b)

Figure 1.11. Feynman graph for Compton scattering, the scattering of a photon on an electron.

annihilated at point Q, where an outgoing electron and photon are created. For a short-lived intermediate state the energy is so uncertain that we cannot speak of violation of energy conservation. In the long-lived final state the conservation of energy (and momentum) must again hold. One calls a short-lived intermediate state, the actual existence of which is incompatible with energy conservation, a *virtual* particle. In Figure 1.11(b), an electron is annihilated and a real photon and a virtual electron are both created at Q, again in apparent violation of energy conservation. But after the annihilation of the virtual electron and a real photon, and the creation of a real electron at S, energy is again conserved.

Both graphs have corresponding well-defined mathematical expressions that are exactly the formulas for Compton scattering derived by Klein and Nishina. These mathematical formulas are the true content of the graphs. The virtual particles only describe the propagation of quantum fields—hence the name *propagator*—and one should never give them the intuitive meaning of real particles. One could also execute the above argument with mass uncertainty instead of energy uncertainty. One then says that the particles are "off the mass shell."

The scattering of two electrons on each other can be easily constructed from the vertex graph of Figure 1.10. Figure 1.12 shows two graphs that contribute to this scattering process. Since here virtual photon lines are attached to different electron lines, one speaks of (virtual) photon exchange. One may say that the electromagnetic interaction between charged particles is mediated by (virtual) photon exchange.

The graphs in Figure 1.12 illustrate the principle of perturbation theory. Figure 1.12(a), with only one internal photon line (one propagator), represents the lowest-order contribution. That is, there is no simpler graph contributing to scattering. Since it contains two vertices, this contribution is proportional to the square of the electron charge, e^2. Figure 1.12(b) represents a higher-order

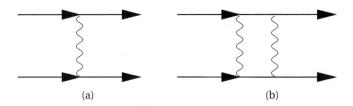

Figure 1.12. Feynman graphs for electron scattering. (a) Lowest-order contribution.
(b) Higher-order contribution.

contribution. It is proportional to e^4. If the interaction is not too strong, then
the higher-order contribution of Figure 1.12(b) is strongly suppressed compared to the lowest-order contribution of Figure 1.12(a), namely by a factor e^2.
A more detailed investigation shows that the suppression of the next-higher
order is typically of order $\alpha = e^2/(4\pi\hbar c)$. This number, called the *fine structure
constant*, is indeed small, approximately $1/137$. Unfortunately, the situation is
not always so simple, as we will see later.

The first field theory in which the idea of field quantization was also applied
to particles with spin $\frac{1}{2}$ (fermions) was the Fermi theory of beta (β) decay, in
which the neutron, generally bound inside the atomic nucleus, decays into a
proton, an electron, and an antineutrino: the neutron is annihilated, while the
proton, electron, and antineutrino are created.

The fundamental interaction in this theory is the product of four fermion
field operators, called the four-fermion interaction. For the beta decay of the
neutron, one has to take the product of the neutrino and neutron field and that
of the Hermitian conjugates of the proton and the electron field, as shown in
Figure 1.13. Since the (nonadjoint) neutrino field contains the creation operator for an antineutrino, this incoming line represents an outgoing antineutrino. With his theory, Fermi obtained excellent agreement with all observed
beta decays. From the life span of the neutron, for instance, one could determine the strength of the four-fermion coupling. This constant, called the Fermi
constant G_F, is very small, and therefore this interaction is called the *weak*
interaction.

After the appearance of Fermi's paper, Heisenberg tried to explain the force
between a proton and a neutron by the exchange of a virtual electron–neutrino
pair. It turned out, however, that at the relevant distances, the resulting force
was much too weak to explain the strong binding inside the atomic nucleus.
The force that binds the nuclei together is called the *strong* interaction.

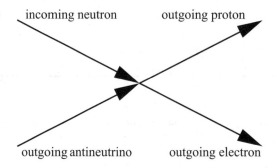

incoming neutron   outgoing proton

outgoing antineutrino   outgoing electron

Figure 1.13. Graphical representation of the beta decay of the neutron through the four-fermion interaction. The outgoing antineutrino created in the decay is represented by an incoming line.

The Japanese physicist Hideki Yukawa was inspired by these considerations, and he proposed to explain the short-range strong interactions inside the nucleus by a hitherto unknown new particle that interacts strongly with the proton and neutron. In the following, we shall refer to it by its current name, *meson*. The interaction is a three-particle interaction analogous to the electromagnetic interaction (Figure 1.12). The electron is replaced by a neutron or proton, the photon by the postulated new meson. The graph is displayed in Figure 1.14. Such a vertex at which a spinless particle couples to fermions is called a *Yukawa coupling*.

The lowest-order proton–neutron interaction is shown in Figure 1.15(a). Using the vocabulary introduced above, this theory explains nuclear forces by meson exchange. The greater the coupling of the meson to the neutron and proton, the stronger the resulting interaction. The range of the interaction is

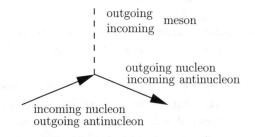

outgoing
incoming   meson

outgoing nucleon
incoming antinucleon

incoming nucleon
outgoing antinucleon

Figure 1.14. The coupling of the meson to the nucleons.

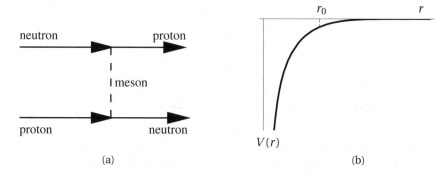

Figure 1.15. Explanation of strong interactions (nuclear forces) by meson exchange. (a) Exchange of a virtual medium-heavy particle between a proton and a neutron. (b) The resulting interaction between the proton and the neutron as a function of the relative distance.

inversely proportional to the mass of the exchanged meson. This important relation between mass and range can be made plausible, at least qualitatively: if the exchanged particle is very heavy, the violation of energy conservation during its exchange is very large, and the virtual particle can live only a very short time and therefore travel only a short distance. The quantitative relation between the mass m of the exchanged meson and the range r_0 of the interaction is given by $r_0 = \hbar/(mc)$.

The interaction energy (potential) between a proton and neutron as a function of distance r is displayed in Figure 1.15(b). It was known that the range r_0 of the nuclear forces is about a femtometer (one millionth of a nanometer), and from that, Yukawa could estimate the mass of the meson to be around $200\,\text{MeV}/c^2$.

I want to emphasize again that the concept of "particle exchange" should not be taken too literally. The graph of Figure 1.15(a) can be interpreted in two ways. According to the first interpretation, at the upper vertex, the incoming neutron is annihilated and a negative virtual meson and a real proton are created; at the lower vertex, the virtual meson and the real proton are annihilated and a real neutron is created. In the alternative interpretation, a proton is annihilated and a virtual positive meson together with a real neutron is created, and the positive virtual meson together with the neutron is annihilated and a real proton created. In the second interpretation, a positive virtual meson runs from down to up, while in the first interpretation, a negative virtual meson runs in the opposite direction. The seemingly contradictory interpretations are

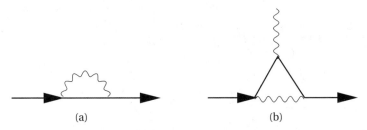

Figure 1.16. Examples of quantum corrections. (a) A contribution to the "self energy" of the electron; the electron interacts with its own radiation field. (b) A quantum correction to the charge of the electron.

compatible if one considers that "virtual particle exchange" is a paraphrase for the propagation of a quantum field, and this quantum field describes the positive meson as well as its negative antiparticle.

In quantum field theory one can calculate the interaction of an electron (or positron) with its own radiation field: an electron emits a virtual photon and absorbs it again, as shown in Figure 1.16(a). Such a contribution is called a *quantum correction*, since it is typical for the occurrence of annihilation and creation operators. If one tries to calculate the corresponding expression, one is unpleasantly surprised: the result of the calculation is infinite, which means that the interaction of the electron with its own radiation field leads to an infinite interaction amplitude and thus to an infinite mass for the electron. In the same way, the correction to the coupling of the photon to the electron, (see Figure 1.16(b)), leads to an infinite coupling strength, and thus to an infinite charge. The reason for these infinities is the appearance of intermediate states with very high energy. Although these states can live only a very short time, there are so many possibilities for them that finally the result diverges and becomes infinite.

We shall meet this problem on several occasions. Therefore, I will very briefly describe a method that circumvents these difficulties. The approach was developed by R. Feynman, J. Schwinger, S.-I. Tomonaga, and F. Dyson in the late 1940s. First, the expressions are *regularized*, which can be accomplished by omitting the contributions with an energy higher than a certain limit, called the *cutoff*. Then the results are finite, but they depend on the arbitrarily introduced cutoff energy. The parameters, such as mass and charge, used in this regularization procedure are called the *bare* parameters. After regularization, the physical parameters are *renormalized*. This is done in the following way: for a certain cutoff one chooses the input parameters in such a way that certain results agree with experiments at a fixed energy. Then one lets the cutoff go to infinity, always insisting on the correct value of the calculated quantity at the fixed energy, called the *scale*. The resulting values for the physical parameters are called *renormalized quantities*. They are now independent of the cutoff, but they will in general depend on the scale. After this procedure has been carried out, one can calculate all graphs at all energies with these parameters, and one obtains well-defined results. A theory for which a finite number of renormalized parameters is sufficient at each stage of the calculation is called a *renormalizable theory*.

Freeman Dyson has shown that QED is a renormalizable theory with two essential experimentally determined parameters: charge and mass. In QED there is a natural choice of the scale for the renormalization procedure: it is the limit of zero energy; that is, for photons the wavelength becomes infinite. In this limit, classical physics is valid, and one can take as renormalized values for charge and mass the values obtained in the era of classical physics, when nothing was known about the subtleties of quantum field theory. Therefore, I have called this choice of the scale "natural." Later, we shall come to know quantum field theories without such a natural choice. We shall learn more about the impressive successes of renormalized perturbation theory in Section 2.4.

In contrast to QED, Fermi's four-fermion theory of weak interactions is not renormalizable. Here one has to introduce more and more parameters as one attempts to calculate to higher and higher order. The four-fermion interaction displayed graphically in Figure 1.13 leads to a six-fermion interaction (see Fig-

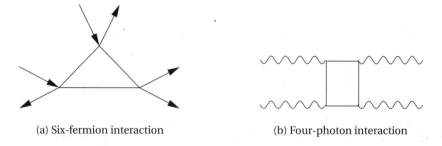

(a) Six-fermion interaction   (b) Four-photon interaction

Figure 1.17. New couplings through quantum corrections. (a) The six-fermion interaction, which follows from the four-fermion interaction (Figure 1.13), cannot be calculated; it is a new parameter in the theory. (b) The four-photon interaction, which follows from the renormalizable electron photon interaction (Figure 1.10), can be calculated.

Figure 1.18. Contributions to the self energy of the vacuum. Here a virtual electron–positron pair is created spontaneously out of nothing (a vacuum).

ure 1.17(a)), which cannot be calculated in the framework of the four-fermion theory. It leads to an infinite result, and therefore a new parameter must be introduced into the theory. It will be the renormalized six-fermion coupling. In contrast, the four-photon interaction, (see Figure 1.17(b)), can be calculated in renormalized QED. It gives a finite result.

One can construct graphs without external lines, that is, without incoming or outgoing particles. Examples of such graphs are displayed in Figure 1.18; they show the creation of a virtual electron–positron pair out of nothing (a vacuum). These graphs lead to an energy density of the vacuum, and—as you may already have suspected—the result is infinite. From the point of view of particle physics, these infinities can be ignored. A vacuum energy does not make itself noticeable, for example, in electromagnetic reactions; think of the enormous rest energy of a particle, $E = mc^2$, which remained unnoticed until 1905. In gravitation, however, such an energy density could be observed, since an energy density contributes to the gravitational force. The vacuum energy in quantum field theory is one of the great unsolved mysteries at the boundary of particle physics and cosmology, to which we shall return in Section 7.6.

1.4.3   Quantum Physics and Errors

Another effect of quantum physics is what are called statistical errors. Of course, errors in measurement existed long before the creation of quantum mechanics, but these seemed only to be determined by some imperfection of the measurement and were in principle reducible to an arbitrarily small value. Quantum mechanics is a deterministic theory, that is, the states change over time following firm laws, but the statements concerning the outcome of an experiment can in general be predicted only with a certain probability. Max Born's recognition of this in 1926 was a major step in the development of modern quantum theory. Einstein did not like this theory, and in a letter to his friend Born, Einstein made the famous statement that he believed that God does not play dice. Among other sometimes paradoxical consequences, a new source of error comes into the mix through this probabilistic nature of quantum physics. These errors are called *statistical* errors, about which nothing can be done other than an extended series of measurements—under certain circumstances too long to be feasible.

In some cases the probability for a certain event that is predicted by quantum mechanics is either zero or one. Then there is, of course, no uncertainty. But in many cases this is not the case. A good example is the lifetime of an unstable state, such as an excited atom, an excited nucleus, or an unstable elementary particle. If we know the underlying dynamics, we can calculate the lifetime with a precision limited only by our capacity of calculation and uncertainty in our knowledge of some fundamental constants.

Let us assume that we have calculated for a certain state a half-life of exactly one second, that is, the probability that it decays within the next second is exactly 50%. If we have only one copy of this state, it is evident that we cannot test this statement very precisely. We can, however, find out whether the statement is likely to be grossly wrong. After 20 seconds, the probability for survival of the state is only 0.0001%, which is close to zero, and if the state has not decayed yet, we should consider the possibility that our theory is wrong— or that we made an error in the calculation. If we have 1,000 copies of the state, then after one second, 500 should have disintegrated: at least that is the most probable number. But due to the probabilistic nature of quantum-mechanical predictions, there can be deviations from this number. It could very well be that 490 or 520 have decayed, even if the theory is correct and the half-life is exactly one second.

The behavior of these variations from the most probable result is described by the standard deviation. A given event is expected to occur with a probability of 66% within the range given by the standard deviation. The standard deviation is approximately equal to the square root of the number of events. In our case we expect 500 events, and the square root of 500 is about 22, and so we expect with a probability of 66% to observe between 478 and 522 decays. But with a probability of 34% there can be more or fewer decays. If an experimentalist tests the theory with 1,000 copies and finds in one second 450 decays, such a result by no means disproves the theory. It is only slightly more than two standard deviations away from the mean value, which occurs with a probability of a few percent. In particle physics one speaks of an *interesting effect* if the deviation from an expected result is more than three standard deviations, and of a *discovery* if the deviation is more than five standard deviations.

In this case, the probability that the deviation is due to chance is only 0.00007%. If one theory predicts a lifetime of 1 second and another one predicts 1.1 seconds, one has to measure the decay of enough copies so that the standard error is only 2%; only then can the experiment discriminate between the two theories on the basis of five standard deviations. If one counts the decays in 1 second, one has to observe in that case about 2,500 events. This experiment can be made more sophisticated, but statistical errors are unavoidable and can be made small only by investigating large numbers of events.

Since in particle physics the production of particles is sometimes very expensive, statistical errors are a serious problem and often set the limits for testing theories.

1.5   Symmetries in Particle Physics

Symmetries have played an important role in physics for a long time, but their significance has increased in particle physics. The great importance of symmetries in particle physics was first recognized by Hermann Weyl and Eugene Wigner. Later, we shall investigate on several occasions why symmetries play such an important role, but first we must examine some rather formal considerations. We will not attempt to use the full highly polished formalism of mathematics related to symmetries, but I hope to convey at least some glimpse—through the keyhole, so to speak—into the wonderful world of group theory.

1.5.1   Symmetries and Transformations

In thinking of symmetry, one's first association may be a rather passive, aesthetic image, but in reality, symmetry becomes apparent only through an action: a symmetry transformation. Think of an apparent example of symmetry such as a butterfly with extended wings. The observer is charmed by finding out that the two wings are congruent. In order to see that this is the case, he has—at least in his mind—to superimpose one wing above the other. The symmetry operation here is reflection through the axis of the butterfly. The manufacturer of a kaleidoscope makes use of several reflections in order to produce a symmetrical pattern. If in the following, we seem to be treating a rather dry subject, namely mathematical transformations, please bear in mind that these operations can disclose beautiful symmetries.

Rotations in space form an especially important group. Here the connection with symmetries is apparent. If we rotate—only in our mind, of course—

Figure 1.19. The rose window of the cathedral in Laon, France. It has 12 equal segments; after a rotation by 360/12 = 30 degrees, it becomes congruent to its original configuration.

the rose window of the Laon Cathedral (Figure 1.19) by an angle of 30 degrees, it comes again into congruence. Its form is thus invariant under a rotation of an integer multiple of 30 degrees. The massive wheel of a prehistoric oxcart comes into congruence after a rotation through an arbitrary angle—if we ignore imperfections in the construction or those due to the ravages of time. In the case of the rose window, the symmetry is called *discrete*, since only certain angles lead to congruence, while in the case of the massive wheel the symmetry is said to be *continuous*.

Rotations in space are intuitively accessible to everybody, and it is convenient that we can use them to study essential features of many symmetry transformations. There are three essential properties of such transformations:

1. Two rotations performed successively result in another rotation. If I spin on my vertical axis and then do a handstand, I have performed two successive rotations (An acrobat could have achieved the same position in a single action). We say that the product of two rotations is again a rotation.

2. We can undo a rotation by reversing the motion. One calls the rotation that undoes another rotation its *inverse*. Thus every rotation has an inverse.

3. The third property seems very trivial, but I must mention it since it is important for abstract mathematical considerations: if one performs three rotations successively, it does not matter whether one groups the first two or the last two into a single rotation.

A set of mathematical objects having these three properties is called a *group*. Therefore, the expression "group of rotations" has also a deeper mathematical meaning. Rotations indeed form a group, with the self-explanatory name *rotational group*.

Another important group of transformations is that of translations: shifts in space and time. These are related to symmetries, too, namely to those of crystals; if a crystal is shifted exactly by the distance of two atoms, it becomes, up to the border regions, congruent with itself. We can then consider translations in space and time as continuous symmetry operations, since space and time are congruent after an arbitrary shift in space or time. This means that the outcome of an experiment is—with all other conditions unchanged—independent of the place where, and of the time when, it was performed. This invariance of physical processes under space and time translations has far-reaching consequences. For example, it leads to conservation of momentum and energy; more precisely, the invariance allows one to define quantities that we call momentum and energy, and these are conserved. The relation between symmetries and conservation laws is a consequence of a mathematical theorem of Emmy Noether.

The group of translations is *commutative*. It does not matter whether I first go five steps in the forward direction and then three steps sideways to the right or if I first go three steps sideways to the right and then five steps forward. In both cases I come to the same place. Commutative groups are called *abelian*, in honor of the Norwegian mathematician Niels Henrik Abel.

The group of rotations, on the other hand, is not commutative. The position I take if I first turn myself by 180 degrees around my spinal axis and then lie down on the floor differs from the position I assume if I first lie down and then turn myself. The groups in which the result depends on the order of the operations are called *noncommutative* or *nonabelian*. Abelian groups have a very different mathematical structure from nonabelian groups, with far-reaching physical consequences, as we will see later.

Before we continue with mathematics, we briefly digress into how symmetries are verified in physics. I have mentioned earlier that the outcome of an experiment does not depend on the place where it is performed, but have added the condition "with all other conditions unchanged." To realize unchanging conditions is sometimes quite difficult, but slightly changed conditions can often be taken into account by calculated corrections.

We all have learned that for small oscillations the period of a pendulum depends only on its length. But a very precise pendulum clock brought in 1672 from Europe to Cayenne, in South America, was a bit slow in its new home. Huygens was able to explain this phenomenon, which at first seems to contradict the homogeneity of space. Near the equator, the gravitational force of the earth is diminished by the centrifugal force due to the rotation of the earth, which in turn leads to an exactly calculable increase of the oscillation period. Since Huygens had made use—at least implicitly—of the homogeneity of space, the exact agreement of his calculation with the observed difference was a corroboration of his hypothesis. A bit more subtle are the temporal fluctuations of certain experimental results at the big particle accelerator at CERN. They can be explained by the tidal forces of the moon, which are strong enough to deform the accelerator in such a way that the energy of the particles changes slightly.

But let us return to rotations, where we still can learn some mathematical concepts. Up to now, we have considered rotations in a purely intuitive geometrical way. But in order to make numerical calculations we have to consider them algebraically. This is done through *group representations*. We can represent each point of a body by three coordinates. If the body is rotated, the coordinates change—not arbitrarily, but in a well-defined way, determined by the particular rotation. This change of coordinates can be represented by a quadratic array of nine numbers, a 3×3 matrix. Figure 1.20 shows a magic square; it is a 4×4 matrix.

I am not going to explain the algebra of matrices. In our context, all that is important is that each rotation can be represented by a matrix and that this possibility of representation by a matrix holds for many groups. Very often, it is possible to classify groups by their simplest matrix representations.

The result of using such representations is that one can introduce rotations purely algebraically, namely, as the group of 3×3 matrices that leave lengths and angles unchanged. It might seem natural to represent rotations in three-dimensional space by 3×3 matrices, but there are also representations by matrices of higher dimension, such as 6×6 matrices. They rotate more complex objects such as fluxes of force or stress tensors. Generally, the size of a matrix determines the number of objects in a *multiplet* that are transformed by the transformation. A $d \times d$ matrix acts on a multiplet of d elements, and one calls

Figure 1.20. Example of a 4 × 4 matrix: the magic square from Dürer's engraving *Melancholia*.

d the *dimension* of the representation. The so-called *fundamental*, representation of rotations in space is of dimension $d = 3$, corresponding to the three coordinates of a point in space.

For each transformation there exists a particularly simple representation, which is called trivial. If a quantity, such as the volume, remains unchanged under a rotation, one can also say that with respect to this quantity all rotations are represented by multiplication by the number 1, and this identity operation is the *trivial representation* of the group. Its representation matrix is a 1 × 1 matrix with only one entry, the number 1. Quantities transforming under the trivial representation, that is, those that do not change at all, play a special role, since they show a particularly high degree of symmetry. If in the following, I speak of the simplest representation, then I mean the simplest nontrivial one.

We have to introduce another important concept, that of generators. In order to uniquely determine a rotation, I need three angles, and conversely, these three angles fix the rotation uniquely. In order to construct a rotation, I therefore need three generators that determine how to construct the rotation with the three given angles. Broadly speaking, the three generators of the rotational group are obtained from rotations by infinitesimally small angles. The fact that rotations are not commutative is reflected in the commutation relations (see Section 1.4.2) of the generators. We denote the generators for rotations around the *x*-, *y*-, and *z*-axes by $\mathbf{L}_x$, $\mathbf{L}_y$, and $\mathbf{L}_z$; all rotations can be constructed out of these three generators. The generators obey rather simple commutation relations, which can be obtained from the geometric properties of rotations. For example, we have

$$\mathbf{L}_x \cdot \mathbf{L}_y - \mathbf{L}_y \cdot \mathbf{L}_x = i\mathbf{L}_z.$$

Here again, $\mathbf{L}_x \cdot \mathbf{L}_y$ means first apply $\mathbf{L}_y$ and then $\mathbf{L}_x$. The commutation relations of the generators nearly determine the group, and in quantum mechanics a miracle occurs that we will treat in the next section.

The concept of generators is not confined to rotations. There is a whole class of groups that can be constructed with the help of generators. These are the so-called Lie groups, which play an important role in physics.

I have consciously used words like "group," "generator," and "represent" that have meaning in everyday language in addition to more technical definitions. However, one should bear in mind that they have a very precise meaning in mathematics.

1.5.2   The Miracle of Spin

One of the essential differences between quantum physics and classical physics is the role of observables; in quantum physics, operators are assigned to observable (measurable) quantities. The result of a measurement is determined by the result of the action of the operator on a state. In quantum mechanics as well as in classical physics, the angular momentum observables play a crucial role. The operators that correspond to these important observables are essentially the generators of the rotational group. This can be derived from the fact that the above-mentioned commutation relations between the generators are the same as those that can be derived for the angular momentum operators from the commutation relations between the position and momentum operators (see Section 1.4.2). This is remarkable, but not yet the miracle.

I have mentioned that the commutation relations do not fully determine the group. By investigating what other groups can be constructed from generators with the same commutation relations, one finds in addition to the group of rotations another one, which can be considered to be rotations in a two-dimensional space. However, this two-dimensional space is not an ordinary plane but one in which the coordinates of a point are complex numbers. This group is called SU(2), where SU stands for *special unitary*, which designates a rotation of points with complex coordinates, and the (2) indicates two dimensions.

These new operators can be called *spin* operators. Thus far, we have defined them only formally. Since the space is two-dimensional, the simplest nontrivial spin has two possibilities for its orientation. In quantum physics, it can be shown that the possible values for the angular momentum are integer multiples of the (reduced) Planck constant $\hbar$, but for the spin, the possible values are multiples of $\hbar/2$. The simplest objects transforming under the spin group SU(2) have two components and therefore are *doublets*; they are called *spinors*. In the following, we will always quote the spin in natural units, so spin $\frac{1}{2}$ means spin $\frac{1}{2}\hbar$, and we say that the spin is a half-integer.

All this is mathematically rigorous, but on the other hand, it sounds a bit like playing around with abstract concepts. But now comes the miracle: to the purely formally introduced spin operators there actually correspond physical observables. There exist particles with half-integer spin values, and these particles are by no means exotic, but constitute the matter surrounding us: electrons, protons, and neutrons.

If a particle has spin $\frac{1}{2}$, its spin can be oriented either parallel or antiparallel to a fixed direction and has with reference to this direction the values $+\frac{1}{2}$ or $-\frac{1}{2}$, normally represented as up, $\uparrow$, and down, $\downarrow$. If such a state is rotated by 180 degrees around an axis perpendicular to the fixed direction mentioned above, the state with spin orientation $+\frac{1}{2}$ ($\uparrow$) changes into spin orientation $-\frac{1}{2}$ ($\downarrow$). This has consequences: if the interaction does not change under rotations, that is, if it is invariant under the rotational group, then the experimental results are invariant too. This means that if a particle with spin $\frac{1}{2}$ is involved, it does not matter whether its spin orientation is $+\frac{1}{2}$ or $-\frac{1}{2}$ ($\uparrow$ or $\downarrow$).

The symmetry properties of a system consisting of two or more particles are of special importance for our further analysis. Let us begin with a system of two particles with spin $\frac{1}{2}$, for which both spins have the same orientation ($\uparrow\uparrow$). Performing a rotation by 180 degrees, we arrive at a state in which both spin orientations are $-\frac{1}{2}$ ($\downarrow\downarrow$). If symmetry under rotations holds, then the interaction between two particles with spin orientation $+\frac{1}{2}$ ($\uparrow\uparrow$) is equal to that for two particles with spin $-\frac{1}{2}$ ($\downarrow\downarrow$). This has been confirmed countless times experimentally. More difficult is the situation in which one particle has spin $+\frac{1}{2}$, the other $-\frac{1}{2}$ ($\uparrow\downarrow$). Here, under rotation by 180 degrees, the state $+\frac{1}{2} - \frac{1}{2}$ ($\uparrow\downarrow$) becomes $-\frac{1}{2} + \frac{1}{2}$ ($\downarrow\uparrow$): the order is reversed. Mathematics tells us that it is useful to form the sum and the difference, that is, the states $s = (\uparrow\downarrow + \downarrow\uparrow)$ and $d = (\uparrow\downarrow - \downarrow\uparrow)$. It turns out that the sum s can be obtained from the state $+\frac{1}{2} + \frac{1}{2}$ ($\uparrow\uparrow$) through a rotation by $90°$, whereas the difference d is invariant under rotations. The four possible combinations ($\uparrow\uparrow$), ($\uparrow\downarrow$), ($\downarrow\uparrow$), and ($\downarrow\downarrow$) can thus be combined into two

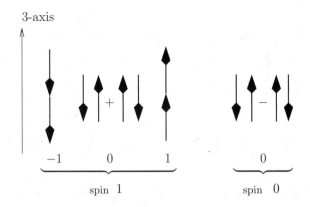

Figure 1.21. Coupling of two spin-$\frac{1}{2}$ particles to total spins 1 and 0.

groups: the triplet $\{(\uparrow\uparrow), (\uparrow\downarrow + \downarrow\uparrow), (\downarrow\downarrow)\}$, the objects of which are transformed into each other, and a singlet $\{\uparrow\downarrow - \downarrow\uparrow\}$, which is invariant under rotations. The triplet has total spin 1 and possible spin orientations $+1, 0, -1$. The singlet with only one possible value has total spin 0; this is displayed graphically in Figure 1.21.

If symmetry under rotations holds, the interaction is independent of the orientation of the spin; that is, it is the same for all three members of the triplet. Symmetry thus reduces the interaction terms possible in principle from four to two, to that of the triplet and that of the singlet.

If we use the language of representation theory introduced in the previous subsection, we may say that spin 1 transforms according to a three-dimensional representation and spin 0 according to the trivial representation.

1.5.3 Isospin

It is quite remarkable that there are physical observables corresponding to these mathematical concepts, which have found a considerable number of technical applications—for example, in medical diagnostics. But there is also a second miracle around the corner: nature has made extensive use of the group SU(2) a second time, as an *internal symmetry*. Internal symmetries refer to properties that have no connection to behavior under space-time transformations. The concept of internal symmetries is one of the most remarkable discoveries of particle physics. Chadwick's discovery of the neutron in 1932 was the beginning of this development (see Section 1.2).

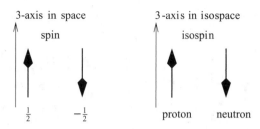

Figure 1.22. The analogy between spin and isospin; the proton has isospin $+\frac{1}{2}$, the neutron $-\frac{1}{2}$.

Immediately after this discovery, Heisenberg guessed that the atomic nucleus is composed of protons and neutrons and not—as assumed previously—of protons and electrons. The masses of proton and neutron are quite similar; the difference is only about one part in a thousand, and from properties of the nucleus one could conclude that the binding forces of neutrons and protons must be quite similar too. Therefore, Heisenberg proposed in 1932 that the proton and neutron are related by a symmetry transformation. In this respect, proton and neutron are different states of one and the same particle, which differ from each other only by their orientation in an internal symmetry space. The analogy with spin was obvious: the electron can also appear in two spin states, and one would not speak of two different particles, but of two different states of one and the same particle. Since 1941, the proton and neutron have been collected together under the name *nucleon*.

Heisenberg introduced the name ρ-spin for the distinctive properties of proton and neutron. Today, the name *isotopic spin*, or *isospin* for short, is in common use. One calls the internal symmetry space *isospace* and says that the proton has isospin orientation $+\frac{1}{2}$, the neutron $-\frac{1}{2}$; see Figure 1.22. Earlier, Heisenberg had used the two-dimensional matrices introduced by Pauli to describe spin; they are the generators of the group SU(2).

However, Heisenberg's formalism was regarded as clumsy and of little use, and was for a few years nearly forgotten. It was revived only in 1936 by B. Cassen and E. Condon, who utilized Heisenberg's formalism to help them to interpret new experiments on charge independence of nuclear forces. They made full use of the formal identity of spin and isospin. As in spherically symmetrical situations, results are unchanged under rotations, so nuclear forces were assumed to be independent of rotations in isospace.

These rotations transform a neutron into a proton, just as rotations in ordinary space transform an electron with spin orientation $-\frac{1}{2}$ into one with orientation $+\frac{1}{2}$. I want to emphasize, however, that the rotation in isospace is a purely formal operation and has no direct intuitive meaning.

Nevertheless, we can transfer the formal findings about the behavior of spins, described above, directly to the behavior of two nucleons. For that purpose, we have only to replace the spin orientation $+\frac{1}{2}$ ($\uparrow$) by a proton, and the orientation $-\frac{1}{2}$ ($\downarrow$) by a neutron. We then go from the situation represented in Figure 1.21 to that represented in Figure 1.23. We thus can combine a pair of nucleons into a triplet with isospin 1 and a singlet with isospin 0.

These considerations encouraged Nicholas Kemmer to go a decisive step further. He attributed this symmetry, which was introduced only to classify states, to the dynamics of meson exchange. It is necessary here to provide some background to Kemmer's hypothesis. Prior to 1937, S. H. Neddermeyer and C. D. Anderson discovered a new kind of particle with a mass of about 240 electron masses and electric charges $+1$ and -1. This particle, with a mass between those of the electron and the nucleon, was at first called a *mesotron*. In 1937, Yukawa's paper on the meson theory (see Section 1.4) of nuclear forces (a theory that Yukawa proposed in 1934) was quoted for the first time in a Western journal. This publication greatly increased interest in the theory, which will be discussed in detail in Chapter 2. J. R. Oppenheimer and R. Serber suggested that the new mesotrons detected by Anderson and Neddermeyer were the particles postulated by Yukawa. Only positively charged and negatively charged mesotrons had been observed, but in order to obtain symmetry, Kemmer had

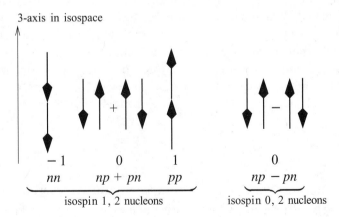

Figure 1.23. Coupling of two nucleons to isospin 1 and 0. The mathematical formalism is identical to that of coupling to spins in Figure 1.21.

to make a daring hypothesis: the existence of a third, neutral, mesotron, which together with the charged ones forms a triplet with isospin 1 (isotriplet). Based on these assumptions, Kemmer could construct an interaction between the isospin doublet of nucleons and the isospin triplet of mesons that was invariant under rotations in isospace. He could prove that this invariance pertains as well to the interaction of nucleons among themselves.

Thus was the first internal symmetry introduced in particle physics, that is, a symmetry that is independent of the space-time behavior of particles. It was modeled after the example of spatial rotations, and the (quantum-mechanical) formalism developed there could be directly applied to the isospin. Today, internal symmetries play a decisive role in particle physics, and we will return to this subject frequently. It was impossible to extend the invariance under rotations in isospace to all interactions, since the electromagnetic interactions of the proton and neutron are very different. Therefore, this invariance was assumed to be valid only for strong interactions, namely, the interactions responsible for the forces inside the atomic nucleus. One says that the isospin symmetry is broken by the electromagnetic interaction.

This breaking of the symmetry was also used as an explanation for the fact that the proton and the neutron do not have exactly the same mass. The small difference in mass should be due to electromagnetic interactions, though it seemed somewhat mysterious that the neutron was heavier than the proton. From classical arguments one would have concluded that the electrostatic repulsion inside the proton makes it heavier than the neutron. The breaking of the symmetry could also explain why one could not detect the neutral mesotron postulated by Kemmer: it could decay by the electromagnetic interaction into two photons. In 1940, S. Sakata and Y. Tanikawa had calculated the mean lifetime of the neutral particle and found a lifetime of about one-hundred-millionth of a nanosecond (10^{-17} seconds). This lifetime is about one hundred billion times smaller than that of the charged mesotrons, and therefore it could not be detected by the methods of Neddermeyer and Anderson.

1.5.4   Discrete Symmetries

Discrete symmetries are not particularly discreet, rather the related transformations do not depend on a continuously varying parameter, as we mentioned at the beginning of the section. The discrete symmetries we will now discuss are the following:

- Space reflection (or *parity transformation*) **P**. In this transformation each point of space is reflected through the origin—that is, the coordinates of a point change their signs.

- Time reversal **T**. Here the direction of time is reversed, like running a movie backward.

- Charge conjugation **C**. Here particles are transformed into antiparticles. For example, an electron is transformed into a positron. This transformation plays a role only in quantum field theory.

Though we have little experience with space reflections through a point, we are all familiar with reflections in a plane from a mirror. You can convince yourself that your right hand in a mirror looks like your left hand in front of the mirror. This is also true for space reflections through a point: a right hand is transformed into a left hand.

In everyday life we have no experience of time-reversal symmetry. Thus growing older is a normal process, while getting younger happens only in science fiction and wishful thinking. Nevertheless, the laws of mechanics and classical field theory are invariant under space reflection and time reversal. At first, this was also assumed to be true in particle physics, but our eyes were opened in the course of time. I will return to this subject later in detail.

One can assign an *internal parity* to a field. If the components of the field change their sign under space reflection, the parity is called "odd" and assigned the value -1. If they stay unchanged, the parity is "even," with the value $+1$. This parity is also assigned to the particles that are the quanta of the field. Under space reflections, the components of the electromagnetic field change sign: thus, photons have parity -1.

There also exist oriented quantities that do not change direction (that is, the sign of the components) under space reflections. They are called *pseudovectors* or *axial vectors*. Spin and angular momentum are examples of axial vectors. The parity of some particles is given in Table 1.1. For particles with half-integer spin, the parity of the antiparticle is opposite to that of the particle. The total parity of a state is the product of the internal parities and the external parity. The latter is determined by the angular momentum. If an interaction is invariant under space reflections, the total parity remains unchanged for all reactions due to this interaction.

Just as parity was defined with respect to space reflections, one can introduce an internal charge parity (or *C*-parity) with respect to charge conjugation **C**. Even if a particle is its own antiparticle, the sign of its quantum field can

particle	P	C	B	L
photon	-1	-1	0	0
proton	$+1^*$		1	0
neutron	$+1$		1	0
electron	$+1^*$		0	1
positron	-1		0	-1
neutrino			0	1
antineutrino			0	-1

*The parity of the proton and the electron is defined to be $+1$.

Table 1.1. Internal parity P, charge parity C, baryon number B, and lepton number L of some particles.

change under charge conjugation. This is the case for photons, and we therefore assign the photon the charge parity -1. Only particles or states that are neutral can have a definite charge parity; for charged particles one can introduce the so-called G-parity, which is a combination of charge conjugation and a rotation in isospace. However, we shall not discuss this here.

If invariance under charge conjugation holds, the dynamical laws are the same for the world and the antiworld, and the product of charge parities is conserved. This means that even if in some reactions particles and antiparticles are annihilated and completely new particles are created, the product of the internal charge parities of the final state is the same as those of the initial state.

In Table 1.1, two further internal quantum numbers are specified: the *baryon number* and the *lepton number*, the introduction of which was compelled by experimentally observed conservation laws. None of the classical symmetries such as conservation of energy, momentum, and charge forbids the decay of a proton into a positron and photons. If those decays were possible with a mean lifetime not much longer than that of the universe, a large part of matter would already have decayed. Since up to now such decay has not been observed, one introduces a new quantum number, the baryon number B.

Mesons, photons, and electrons have baryon number 0; nucleons (proton and neutron) have $B = 1$. Lepton number L is formed analogously; it is $+1$ for electrons and neutrinos and -1 for their antiparticles. If a neutron decays into a proton, electron, and antineutrino, both baryon and lepton number are conserved: in the initial and final states we have $B = 1$ and $L = 0$. In the above-mentioned unobserved decay of a proton into a positron and photons, the baryon number would change from $B = 1$ to 0, and hence it would not be conserved. We will return to speculations on baryon number violation in Section 7.3.

1.6 The Discovery of the Positron and the Mesotron

Protons, neutrons, and electrons are the building blocks of normal matter, which surrounds us and of which we ourselves consist. The photon plays a decisive role by mediating the electromagnetic interaction. It is by far the most abundant normal particle in the universe. The particles whose discovery is the subject of this section are important for our understanding of the structure of matter, but they do not occur in appreciable amounts. The only natural—that is, not manmade—processes in which they occur are reactions with highly energetic particles from outer space.

During early investigations into the phenomenon of radioactivity, it was already known that ionization chambers lose their charge even if no radioactive material is nearby. This was an indication that air was a conducting material even without apparent sources for ionization. Beginning in 1911, V. F. Hess, in Vienna, investigated the strength of ionization in a series of manned balloon flights and discovered that ionization increases with height. He proposed two explanations: either there is a previously unknown substance occurring mainly at high altitudes, or there is a penetrating ionizing radiation coming from outer space. The evidence for an extraterrestrial origin grew continually stronger, and R. A. Millikan, not only an excellent physicist but also a great lobbyist for science, coined the impressive name *cosmic rays*. Millikan, best known for his precise determination of the elementary charge in his oil droplet experiment, also made significant contributions to the exploration of cosmic rays. It was a subject of intensive research in the 1920s and 1930s, and soon some of the essential properties of cosmic rays became known. In this section, I will report on two important related discoveries: the detection of positrons and of mesotrons.

C. D. Anderson, a student of Millikan, investigated cosmic rays in a cloud chamber placed in a strong magnetic field. A magnetic field exercises a force on a moving charged particle. This force is perpendicular to the magnetic field and the direction of motion, and therefore the trajectory of a particle is curved. The curvature depends on the momentum of the particle: the smaller the momentum, the greater the curvature. Therefore, one can determine the momentum of a charged particle very precisely from the curvature of its trajectory. If

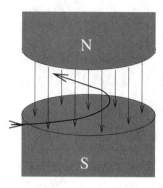

Figure 1.24. The trajectory of a positively charged particle in a magnetic field.

the direction of the motion is known, one can also determine the sign of the charge: a positive charge in a magnetic field that points downward is deflected to the left, while a negative charge is deflected to the right. Figure 1.24 shows the trajectory of a positively charged particle.

Anderson detected many trajectories that corresponded to particles moving from bottom to top—if they were interpreted as electrons (negatively charged). This seemed strange, since most cosmic rays should come from above. In order to determine the direction of the particles, Anderson invented a simple but ingenious device. He put a plate of lead into the middle of the chamber. Since particles can only lose but not gain momentum in this absorber and since the curvature is inversely proportional to the momentum, one could be certain that the particle had moved from the part with lower curvature to the part with higher curvature. In 1931, Anderson found trajectories of the type shown in Figure 1.25.

This particle was definitely moving from the lower to the upper part of the chamber—the curvature is much bigger in the latter than in the former, and hence it has lost momentum in the absorber. In its direction of flight the trajectory was deflected to the left, and from that one could conclude that the charge was positive. From the large penetrating power of the particle, Anderson could draw the equally compelling conclusion that it could not be a proton but must be a much lighter particle. The penetration power of a proton could easily be calculated, since from the curvature one could obtain the momentum and consequently the velocity.

For a given momentum a particle with a smaller mass has a higher velocity and therefore can travel a longer distance in the chamber. One could even con-

Figure 1.25. The trace of a charged particle in a cloud chamber with a lead absorber of thickness 6 mm. From the curvature one can conclude that the particle has moved from the lower part into the upper part, since the curvature is greater in the latter. The north pole of the magnet was above the chamber; thus the magnetic field lines came from above the plane; from that and the deflection to the left one could conclude that the particle was positively charged. It could not be a proton, since a proton of this momentum could not have traveled such a large distance; indeed, it could have traveled at most one-tenth the distance.

clude that the mass of the particle in question was less than 20 electron masses. This was of course a strong indication that the observed trajectory came from the antiparticle predicted by Dirac, the positron, which had met with so little acceptance from so many physicists.

This supposition was proved later by very precise determinations of the mass of the positron. P. M. S. Blackett and G. P. S. Occhialini found pairs of electrons and positrons in their cloud chamber with apparently a common origin. With that was also established the creation of an electron–positron pair, as predicted by Dirac's theory.

This was of course a great triumph of theory. What had seemed to be an unfortunate side effect of an otherwise impressive formalism turned out to be the impressive prediction of a completely new kind of matter. Nevertheless, the discoverers denied having been influenced by Dirac's theory. Anderson said, "Yes, I knew about the Dirac theory .... But I was not familiar in detail with

Dirac's work. I was too busy operating this piece of equipment to have much time to read his papers .... [Their] highly esoteric character was apparently not in tune with most of scientific thinking of the day .... The discovery of the positron was wholly accidental." Blackett and Occhialini were in Cambridge, in direct contact with Dirac. But Blackett said in 1962 that at the time of the discovery of the positron nobody took Dirac's theory seriously.

Such statements like those of Anderson should be viewed with some skepticism. Evidently, many experimentalists of that time thought it more honorable to discover something by chance than to confirm a theory. However, one can say with certainty that in this case, theory did not guide experiment; rather, experiment saved a theory with seemingly absurd consequences. This is in contrast to many later experimental discoveries that were possible only because of very precise theoretical predictions.

Soon after the discovery of positrons in cosmic rays, I. Joliot-Curie and her husband, F. Joliot, discovered that positrons are also emitted in the radioactive decay of certain artificially produced elements. Here a proton in the nucleus decays into three particles: a neutron, a positron, and a neutrino.

Out of this esoteric theory a technique was developed for medical diagnosis: *positron emission tomography* (PET). One injects a radioactive substance that is a positron emitter into the blood. If a positron is slowed down, it catches an electron and is annihilated into two highly energetic photons (*gamma* (γ) *quanta*). These can be detected, and the position where the annihilation takes place can be determined. In this way one can identify the location of an excessive blood supply. Figure 1.26 shows the Feynman graph for the annihilation of a positron and an electron into two photons.

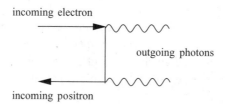

Figure 1.26. Feynman graph for the annihilation of an electron and its antiparticle, the positron. From the released energy of more than 1 MeV two highly energetic photons (gamma quanta) are created. Since the positron is an antiparticle, it is represented by an outgoing arrow, although it is incoming. This effect is used in positron emission tomography (PET).

Progress in the investigation of cosmic rays continued to be made. Using the coincidence method, Bothe and Kohlhörster discovered that there is a component that easily penetrates thick layers of matter, leading to a distinction between hard (penetrating) and soft components. In experiments, the hard component normally produced pictures with single tracks. In the soft component, many tracks passed simultaneously through the cloud chamber. These were called *showers*. In 1934, Hans Bethe and Walter Heitler calculated the energy loss of highly energetic particles using QED, which was still in its infancy. Their theoretical results made an analysis of both components possible: the soft component could be explained very satisfactorily as being composed of electrons and positrons, but the hard component seemed to contradict the theory of Bethe and Heitler.

Anderson and Neddermeyer continued to use their successful detection and analysis device: a cloud chamber in a strong magnetic field with an absorber in it. In 1936, they concluded that either the theory breaks down at high energies—a view not unpopular in those times—or "there exist particles of unit charge but with a mass ... larger than that of a normal free electron and much smaller than that of a proton." In 1937, their statement was more emphatic: either highly energetic electrons have an additional still unknown property or there is a further particle with a mass that is large compared to that of the electron and much smaller than that of the proton. But since no properties were known other than charge and mass, they judged the second hypothesis, that of a new particle, to be more plausible. Their conclusion was founded on the results shown in Figure 1.27.

Charged particles of the same energy were split into two groups. The energy loss in an absorber—a platinum plate in the chamber—was very different in the two cases. In one group, mainly particles from the showers, the energy loss was proportional to the energy, as predicted by the theory for electrons and positrons, but in the other group, that of the single tracks, the energy loss scattered widely and was on average independent of the energy. This was incompatible with the hypothesis that these tracks came from electrons or positrons—if the theory of Bethe and Heitler was correct. Since on the other hand, there were many electrons of high energy for which the loss agreed with theory, it was implausible that the theory was wrong, and therefore they came to the conclusion that they had found a new particle.

The following year, Anderson and Neddermeyer took a decisive picture: a particle passed through a Geiger counter and came to rest in the chamber. From the track one could derive the mass of the particle rather reliably at about

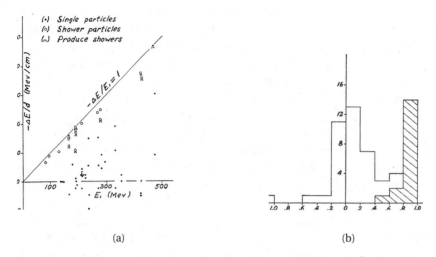

(a)   (b)

Figure 1.27. (a) The energy loss of cosmic rays in a platinum absorber of thickness 1 cm in a cloud chamber. Displayed is the specific energy loss in the absorber, $-\Delta E/d$ (MeV/cm), versus the energy of the particle E_1 (MeV). The open circles represent tracks from a shower; the dots represent single tracks. One can see that only for particles from the showers (open circles) is the energy loss $-\Delta E$ approximately proportional to the energy ($-\Delta E/E_1 = 1$). (b) Relative energy loss $-\Delta E/E_1$; particles from showers are shown in the cross-hatched region. (These figures are reproduced from the original publication of Anderson and Neddermayer.)

240 electron masses. The modern value has been established at 105.658 MeV/c^2, that is, about 207 electron masses.

Theory was extremely important for the analysis of the experiments, but for the discovery of the new particle its influence was certainly weaker even than in the case of the positron. The papers of Yukawa, published in 1934, were first quoted in a Western journal after the results of Anderson and Neddermeyer were known. Nevertheless, it is remarkable that in 1933, P. Kunze, of Rostock, published a picture from a cloud chamber in a magnetic field with a strange track. While stating that this track ionizes too little for a proton and too much for a positron, he comments: "The nature of this particle is unknown." It was his supposition that it comes from an "explosion of the nucleus." However, this result remained unnoticed, perhaps because in 1933 there was no great speculation about a new medium-heavy particle.

I already have mentioned that the discovery of a medium-heavy particle, called the mesotron, gave great impetus to the theory of Yukawa, who had proposed such a particle in order to explain nuclear forces. It seemed plausible that the particle discovered by Anderson and Neddermeyer was the particle postulated by Yukawa. But the more one learned about the properties of the mesotron, the more difficult it became to identify it with the predicted meson. In addition to other indications, the following one turned out to be decisive: if the mesotron is responsible for the strong nuclear forces, it has to interact strongly with the nucleus and to be absorbed by it. For positive particles this is prevented by the electric repulsion through the (positive) nucleus, but for negative mesotrons this effect should be observed. The decisive experiment was performed in Rome in 1946. H. Conversi, G. Pancini, and V. Piccioni were able to show that the negative mesotron is not absorbed by carbon nuclei. In the meantime, however, there was a new willingness to take the idea of previously unknown particles seriously. H. Bethe and R. Marshak as well as S. Sakata and T. Inoue proposed that there are two kinds of mesons: the experimentally observed mesotron of Anderson and Neddermeyer, and Yukawa's meson, still to be discovered. Independently of these speculations the second meson was soon found, but this was already at the beginning of modern particle physics.

1.7 Early Accelerators

To make progress in particle physics, it was critical to be able to accelerate charged particles to high energies in the laboratory, and to no longer have to rely on rare events in cosmic rays. Here I briefly describe the first particle accelerators. In principle, the Lenard tube was a particle accelerator. In it, electrons were accelerated by the high voltage applied at the anode, and the electron beam was extracted in order to perform experiments. A television tube is also an accelerator; the electrons are not only accelerated in the tube, but also deflected and directed to a specific location on the screen.

J. D. Cockcroft and E. T. S. Walton developed an accelerator in which charged particles were accelerated by the extremely high voltage (according to the standards of that time) of 600,000 volts. The resulting proton beam thus had an energy of 600,000 eV. This energy was sufficient for some protons to penetrate a lithium nucleus and split it; this was the first manmade nuclear fission.

Of much greater importance for particle physics was the construction of circular accelerators—the cyclotrons—by E. O. Lawrence, the American physicist. Charged particle are accelerated in these machines not only once by a high voltage, but many times in succession. A charged particle that moves perpendicularly to a homogeneous magnetic field is forced into a circular trajectory;

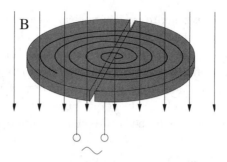

Figure 1.28. Design principle of the cyclotron. The lines of the magnetic field are perpendicular to the camembert boxes, called the D's, which are open at one side. Between the two D's an alternating voltage is applied that accelerates the particles during the transition from one D to the other.

see Figure 1.28. If the velocity of the particle is small compared to the velocity of light, the time needed to travel around a full circle is independent of the particle's energy. Qualitatively, this is easy to understand: if the velocity is high, the circle is large; if the velocity is low, the circle is small. A particle with high velocity (and therefore high energy) has to travel a longer distance to return than does a particle with low velocity, and this leads to the same period for the two cases.

In the cyclotron, the particles move perpendicularly to the homogeneous magnetic field in two metallic D-shaped boxes, called D's, that are open at one side. Between the two D's an alternating voltage is applied with an amplitude of several thousand volts. The frequency of the alternating voltage is adjusted in such a way that a particle is accelerated at each passage from one D to the other. Since this happens twice per revolution, the particles are accelerated many times and can collect a large amount of energy.

If the energy becomes very high, the velocity of the accelerated particle is no longer small compared to the velocity of light, and one has to take into account the effects of special relativity. This is the case if the kinetic energy becomes comparable with the rest energy. For the proton, the latter is 938 MeV, and therefore special relativity has to be taken into account for energies above approximately 100 MeV. Then the period of the moving particles depends more and more on their energy. According to the theory of special relativity, the speed of a particle cannot be greater than that of light. The radius continues to grow with energy, and therefore highly energetic particles take a longer time to travel through a full circle than those with low energy.

The frequency of the alternating voltage between the D's has to be adjusted with increasing energy of the accelerated particles in order to give the particles a "kick" at just the right moment. Of help here is phase stability, which was discovered independently by V. I. Veksler in 1944 and E. M. McMillan in 1945. Particles that arrive too early are accelerated less than those that are just in time (synchronous), and those that are too late gain more energy. In this way particles accumulate in bunches around the synchronous ones. Lawrence constructed a synchrocyclotron with a diameter of 184 inches (4.5 m). On November 1, 1946, a beam of helium nuclei (alpha particles) was accelerated for the first time to an energy of 350 MeV. This marks the beginning of modern accelerator technology. Its further rapid development will be described in Sections 2.6 and 5.6.

2

The Great Leap Forward

After the Second World War, the physics of elementary particles made a great leap forward. Many new particles were detected, some of which had been predicted, while others were unexpected. Quantum field theory was refined, making very precise predictions possible .

2.1 The Predicted Meson Is Found

Somewhat arbitrarily, I identify the beginning of modern particle physics with the discovery of the meson, which was predicted in 1934 by Yukawa. The beginning of modern particle physics is characterized by the last key contributions of cosmic ray physics and the subsequent rapid takeover of the experimental domain by accelerators.

We saw in Section 1.6 that the absorption experiments performed in Rome in 1946 proved definitively that the mesotron of Neddermeyer and Anderson was not the particle predicted by Yukawa to explain the nuclear forces. But the theory was soon rescued. In Bristol, C. F. Powell and his group perfected an old detection method, that of photographic emulsions (see Section 1.3). Cosmic rays contain highly energetic charged particles that can interact with the nuclei of the emulsion and create new particles. Powell and his collaborators searched for such particles. They exposed photographic plates at high altitudes, since the radiation there is less attenuated by the atmosphere than at sea level, and so there are likely to be more reactions. The particles produced in these reactions are detected by the tracks they leave in the emulsion. To look for them is a very time-consuming procedure, since the emulsions have to be scanned laboriously under the microscope for traces.

In photographic plates that were exposed for a month in early 1947 on the Pic du Midi, in the Pyrenees, the scanner Marietta Kurz detected an exceptional trace with a kink. It is shown in Figure 2.1.

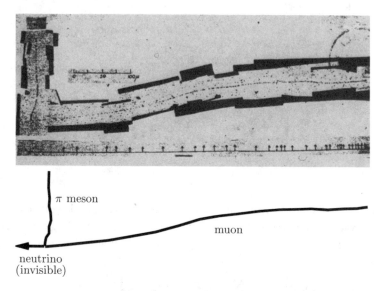

Figure 2.1. The first published image of the decay of a pi meson into a muon. The pi meson enters from above and decays into an apparently lighter charged particle (muon) and a neutral particle (neutrino, invisible). The traces are not straight lines, because the particles are scattered by the nuclei in the emulsion.

This trace was immediately interpreted as evidence for the decay of a medium-heavy charged particle into a lighter charged particle and a very light neutral particle. Confusion with a proton was excluded, since protons create traces that are straighter and more evenly ionized. Soon, a second similar trace was found, and in April 1947, C. G. M. Lattes, H. Muirhead, G. P. S. Occhialini, and C. F. Powell announced the detection of a new medium-heavy particle, the meson. A theory with two mesons already existed, and the following interpretation seemed natural: the decaying heavier particle is the meson proposed by Yukawa; the charged decay product is the mesotron of Neddermeyer and Anderson, and the invisible neutral particle is the neutrino.

A second set of photographic plates was exposed to cosmic rays at an altitude of 5,600 meters near La Paz, in Bolivia. In these plates, 30 traces were detected that could be interpreted as the decay of a meson into a mesotron. In the following, we will call the particle predicted by Yukawa and found by Powell and collaborators the *pi meson* (π-meson), and the mesotron of Neddermeyer and Anderson the *muon*, even when referring to events before October 1947, when these names were proposed.

The detection of pi mesons in cosmic rays was a missed opportunity for accelerator physics. Since 1946, the 184-inch synchrocyclotron had been operating with a beam of alpha particles with an energy of 390 MeV. This energy was sufficient to create pi mesons. But Lawrence's laboratory was more focused on developing accelerators than on performing experiments. When asked by Edward Teller whether the machine at the laboratory was producing pi mesons, the theoretician Robert Serber, who was working at Berkeley, replied, "Sure, but nobody knows yet how to find them." Finally, when Cesare Lattes arrived, bringing the experience of the Bristol group, he was able to find plenty of pi mesons in photographic plates exposed to the beam of the synchrocyclotron. Though the number was larger than in the cosmic ray experiments, the determination of the mass, among other quantities, was less precise. It was some time before accelerators took the uncontested lead in the hunt for new particles.

Nevertheless, it was only a short while before the first new particle was detected with the aid of accelerators. In 1949, an anomalous quantity of highly energetic photons was detected at the Berkeley synchrocyclotron. Since quantum field theory had become highly developed in the meantime, the discrepancy with theory was now taken seriously, and it appeared that these photons were in fact the decay products of a neutral particle of about 300 electron masses. Such a photon could well be the neutral partner of the charged pi mesons. It had been predicted in 1938 by Kemmer to explain the charge independence of nuclear forces (see Section 1.5.3). Shortly thereafter, in 1950, J. Steinberger, W. K. M. Panofsky, and J. S. Steller observed the two photons that appear in the decay together and are detected by simultaneous measurement. The mass of the neutral particle was computed to be (approximately) equal to the mass of the charged pi meson; its lifetime had to be shorter than one-hundred-thousandth of a nanosecond (10^{-14} seconds), in agreement with theoretical predictions. Soon neutral pi mesons were detected in cosmic rays as well.

Strong absorption by light nuclei, which was expected for Yukawa particles (pi mesons), was soon observed, and there was a great euphoria: it seemed that nuclear interactions were now understood just as well as electromagnetic ones. The rapid awarding of Nobel Prizes—1949 for H. Yukawa and 1950 for C. F. Powell—indicates the importance with which these results were viewed.

However, the general elation was soon dampened. It turned out that while a field theory with mesons could make some very important general predictions, it did not allow detailed quantitative statements such as those in the quantum theory of electrons and photons. In the meantime, this theory had become an indispensable tool for analyzing new experiments. The old mesotron, which was initially called a *mu meson* (μ-meson) and later a muon, was dethroned as a theoretically important particle, and I. I. Rabi posed his famous question, "Who ordered that?"

Anderson and collaborators made an important discovery that clarified the muon's properties. In 1947, they observed in a very sensitive cloud chamber the decay of a negatively charged muon into an electron with rather low energy (25 MeV). There were two possible interpretations: they had observed either a decay into an electron and an additional new neutral meson, or a decay into an electron and two neutral particles. Pontecorvo had proposed earlier that with the absorption of a muon by a nucleus, a neutrino is emitted. This means that the muon has spin one-half and is coupled to the neutrino like the electron. Further measurements of muon decays and analysis of the energy distribution of the decay electrons corroborated the hypothesis that the muon decays into an electron and two neutrinos.

The expected decay of the pi meson into an electron and a neutrino had not (yet) been observed, but more precise determinations of the mass made it very probable that a pi meson decays into a muon and a neutrino. The structure of weak interactions as proposed by Fermi (see Figure 1.13) seemed to be universal: for each electron or muon that is created, an antineutrino has to be created too, where the creation of a particle and the annihilation of an antiparticle are considered formally equivalent.

Since 1946, electrons, muons, and neutrinos have been called *leptons*. If one assigns to these particles the lepton number +1 and to the corresponding antiparticles—the positron, positively charged muon, and antineutrino—the lepton number −1, the relations mentioned above can be summarized by a conservation law, namely the conservation of lepton number. This was briefly mentioned in Section 1.5.4.

2.2   Strange Particles Cause Excitement

Whereas the discovery of pi mesons fit very well into the theoretical landscape, in 1946 and 1947 particles were detected that again "nobody had ordered." Physicists did not react with the enthusiasm that had accompanied the discovery of the pi meson, and it was several years before the importance of these discoveries was fully acknowledged, at which time serious effort was made to incorporate these particles into the theoretical framework.

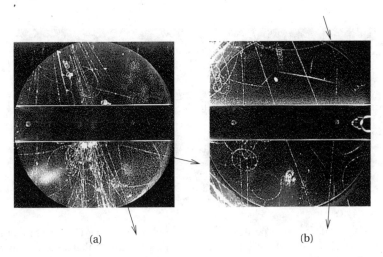

(a)   (b)

Figure 2.2. The first pictures of V particles. (a) The decay of the uncharged V^0 particle into two charged particles takes place immediately below the absorbing plate; the traces form an inverted V (*arrows*). (b) A charged particle enters the chamber at the upper-right corner and decays into another very fast charged particle that flies straight down and passes the absorber. The kink shows that a further invisible (neutral) particle has to occur in the decay.

It all started in Manchester. There, the focus had been particularly on *penetrating showers*—cosmic ray events in which many particles pass easily through thick absorbers. Today, we would call them nuclear reactions induced by highly energetic cosmic rays. C. C. Butler and G. D. Rochester triggered their cloud chamber with such penetrating showers. That is, they took pictures only during times when such showers occurred. In 1946–1947 they obtained two pictures (Figure 2.2) that they interpreted as traces of decays of unknown particles. They interpreted the first one, a slightly oblique inverted V, as the decay of a neutral particle into at least two charged ones. The other one, a kink, looking like a very open V, they interpreted as the decay of a charged particle into another charged particle and at least one additional neutral particle.

Whatever the decay products may have been, the masses of the neutral and charged decaying particles were determined to be between 800 electron masses (400 MeV/c^2) and a value distinctly below the proton mass.

The following two years were frustrating for the Manchester group: they found no further V events. Finally, in 1959, a new and beautiful event was observed. The observation was made by the group in Bristol, which had considerably refined the emulsion technique. The picture is shown in Figure 2.3.

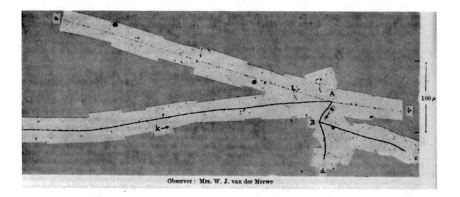

Observer : Mrs. W. J. van der Merwe

Figure 2.3. The decay of a V particle (k). At the point A the particle (k) decays into three charged particles: one of these is a strongly ionizing slow particle that induces a nuclear reaction at point B. The traces of the reaction products lead to c and d. The others are two fast charged particles with lower ionization density that lead to a and b.

The interpretation is clear: A heavy meson, marked by k in the picture, penetrates into the emulsion from the left. It decays at point A into three charged particles: the first is a strongly ionizing slow particle that induces a nuclear reaction at point B; the thick traces of the reaction products lead to c and d. The other two are fast charged particles with lower ionization density that lead to a and b. The slowly decaying particle is a pi meson; the two fast particles must be heavier than electrons. Assuming them to be pi mesons as well, one obtains for the heavy meson (k) a mass of 490 MeV/c^2. Indeed, this interpretation turned out to be correct: a charged K meson with mass 493.677 MeV/c^2 decays into three charged pi mesons. In 1950, Anderson and collaborators found 34 further V-shaped traces. Blackett served as godfather and gave them the name "V particles."

In the meantime, the Manchester group had transported their cloud chamber to the Pic du Midi. After six months, they found 36 neutral and 7 charged V particles. With those events a decay analysis was possible, which yielded the following results: one group of neutral V particles (V_2^0) most probably decayed into a proton and a negatively charged pi meson; the other group (V_1^0) most probably decayed into two pi mesons. Since 1953, the particles that have a proton or a neutron as decay product have been called *hyperons*, and the lighter ones are called K mesons, to distinguish them from the pi mesons.

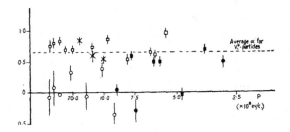

Figure 2.4. From this diagram the Manchester group concluded that there exist two kinds of V particles.

The diagram that led the authors to conclude that there are two kinds of V particles is displayed in Figure 2.4. I will not discuss details, but only illustrate the rather insecure starting point of V-particle physics. Over the decade of the 1950s the fog cleared, and it turned out that there are even more kinds of new particles. The hyperons are divided into three classes, with different masses and decay rates. They were called *lambda hyperons* (Λ-hyperons), *sigma hyperons* (Σ-hyperons), and *Xi hyperons* (Ξ-hyperons). Since 1954, hyperons and nucleons have been collected under the name *baryons* and assigned the baryon number $+1$, while the antiparticles have value -1. The decays are constrained by conservation of baryon number (see Section 1.5.4).

In 1957, Murray Gell-Mann and Arthur Rosenfeld summarized the known properties of elementary particles in a review article in the journal *Annual Review of Nuclear Science*. The essential content is given in Tables 2.1 and 2.2. These tables are of particular historical interest. They gave rise to the "Rosenfeld tables of elementary particles," which at the time were compact enough to be carried in a wallet. Gradually, however, they grew into the *Review of Particle Physics*, a book that appears every two years and now comprises more than one thousand pages. They are accessible online at http://pdg.lbl.gov. The tables explain the initial uncertainty regarding the decays: not only was it difficult to detect the decay particles, but the same type of particle could decay in a variety of ways.

The V particles were the impetus for a number of developments in the theory of elementary particles. Though they were neglected in the initial euphoria surrounding the predicted pi mesons, soon physicists became intrigued by a mysterious property. From their relatively long lifetime, one could deduce that the V particles decay via the weak interaction. On the other hand, they are

	particle	spin	mass (MeV)	mean life (sec)
photons	γ	1	0	stable
leptons & antileptons	$v, \bar{v}$	$\frac{1}{2}$	0	stable
	e^-, e^+	$\frac{1}{2}$	0.510976	stable
	μ^-, μ^+	$\frac{1}{2}$	105.70±0.06	$(2.2 \pm 0.02) \cdot 10^{-6}$
mesons	$\pi^{\pm}$	0	139.63±0.06	$(2.56 \pm 0,05) \cdot 10^{-8}$
	π^0	0	135.04±0.16	$(0 < \tau < 0.4) \cdot 10^{-15}$
	$K^{\pm}$	0	494.0 ±0.20	$(1.224 \pm 0.13) \cdot 10^{-8}$
	K^0	0	493 ±5	$K_1 : (0.95 \pm 0.08) \cdot 10^{-10}$ $K_2 : (3 < \tau < 100) \cdot 10^{-8}$
baryons	p	$\frac{1}{2}$	938.213±0.01	stable
	n	$\frac{1}{2}$	939.506±0.01	$(1.04 \pm 0.13) \cdot 10^3$
	Λ	$\frac{1}{2}?$	1115.2 ±0.13	$(2.77 \pm 0.15) \cdot 10^{-10}$
	Σ^+	$\frac{1}{2}?$	1189.3 ±0.35	$(0.78 \pm 0.074) \cdot 10^{-10}$
	Σ^-	$\frac{1}{2}?$	1196.4 ±0.5	$(1.58 \pm 0.17) \cdot 10^{-10}$
	Σ^0	$\frac{1}{2}?$	1188.3 ±2	$(< 0.1) \cdot 10^{-10}$ theoretically $\sim 10^{-19}$
	Ξ^-	?	1321 ±3.5	$(4.6 < \tau < 200) \cdot 10^{-10}$
	Ξ^0	?	?	?

Table 2.1. Masses and lifetimes of elementary particles, according to the table published by Gell-Mann and Rosenfeld in 1957.

produced so abundantly that one had to assume their creation through the strong interaction. This finally led Gell-Mann, Abraham Pais, and K. Nishijima to assign a new property to the V particles that was called *strangeness*. In many respects, strangeness is similar to electric charge, but in contrast, it is conserved only in strong and electromagnetic interactions, not in weak interactions. Strangeness is conserved in the creation process through the strong interaction, and for each particle with strangeness +1, a particle with strangeness −1 must be produced, for instance a corresponding antiparticle. In the (weak) decay, however, the strangeness can change; that is, the decay products of a strange particle do not have to be strange.

There are weak decays of strange particles in which no leptons occur, and therefore cannot be described directly by Fermi's theory. However, the particles' lifetimes show that the strength of the nonleptonic interactions is similar to that of the leptonic ones. The K^+ meson decays with comparable proba-

	K^+ decay mode			K^0 decay mode	
	branch. ratio (%)	decay. (numb./sec)		branch. (%)	decay. (numb/sec)
identified					
$(\theta^+) \to \pi^+ + \pi^0$	25.6 ± 1.7	$20.9 \cdot 10^6$	$K_0^1 \to \pi^+ + \pi^-$	86 ± 6	$0.9 \cdot 10^{10}$
			$K_1^0 \to \pi^0 + \pi^0$	14 ± 6	$0.15 \cdot 10^{10}$
				100	$1.05 \cdot 10^{10}$
$(\tau) \to \pi^+ + \pi^+ + \pi^-$	5.66 ± 0.30	$4.62 \cdot 10^6$	$K_2^0 \to \pi^+ + \pi^- + \pi^0$	~ 30	$\sim 3 \cdot 10^6$
$(\tau') \to \pi^+ + \pi^0 + \pi^0$	1.70 ± 0.32	$1.38 \cdot 10^6$			
$(K_{\mu 2}) \to \mu^+ + \nu$	58.8 ± 2.0	$48.0 \cdot 10^6$	no analogous decay		
$(K_\beta) \to e^+ + \nu + \pi^0$	4.19 ± 0.42	$3.42 \cdot 10^6$	$K_2^0 \to e^\pm + \nu + \pi^\mp$	~ 30	$\sim 3 \cdot 10^6$
$(K_{\mu 3}) \to \mu^+ + \nu + \pi^0$	4.0 ± 0.77	$3.26 \cdot 10^6$	$K_2^0 \to \mu^\pm + \nu + \pi^\mp$	~ 30	$\sim 3 \cdot 10^6$
	100	$81.6 \cdot 10^6$		100	$\sim 10 \cdot 10^6$
missing (perhaps strictly forbidden)		*forbidden because:*			*forbidden because:*
$K^+ \to \pi^+ + \gamma$	< 2.0	spinless K	$K_2^0 \to \pi^+ \pi^-$	< 5	CP-invar.
			$K_{1,2}^0 \to \pi^0 + \gamma$	< 10	spinless K
$K^+ \to \nu + \nu + \pi^+$	< 2	ν-ν pair emiss.	$K_{1,2}^0 \to \nu + \nu$	< 10	spinless K
$K^+ \to \mu^+ + e^- + \pi^+$	< 0.01	μ-e-pair emiss.	$K_{1,2}^0 \to e^\pm + \mu^\mp$	< 5	μ-e-pair emiss.
etc.					
missing (rare, but perhaps allowed)					
$K^+ \to e^+ + \nu$	< 1.0		$K_{1,2}^0 \to \pi^\pm + \pi^0 + \mu^\mp$		
			etc.		
$K^+ \to \pi^+ + \pi^- + \mu^+ + \nu$	< 0.01		$K_1^0 \to e^\pm + \nu + \pi^\mp$		
etc			etc		

Table 2.2. Branching ratios and decay rates of K mesons according to the review article of Gell-Mann and Rosenfeld as in Table 2.1.

bility into a positive muon (antimuon) and a neutrino, and into nonleptonic products, namely one positive and one negative pi meson. The mean life of a charged pi meson is 26 nanoseconds, while the mean life of a neutral pi meson that can decay via electromagnetic interactions is shorter by a factor of about a billion. In decays that can proceed through the strong interaction, the mean lifetime is shorter by a factor of about another billion.

The relationship between strangeness and charge becomes apparent in a simple equation that holds between the charge Q, the projection of the isospin I_3, the baryon number B, and the strangeness S:

$$Q = I_3 + \tfrac{1}{2}B + \tfrac{1}{2}S.$$

One may easily convince oneself that this relation holds for nucleons and pi mesons. It was discovered independently by Gell-Mann and Nishijima in 1955.

Since isospin is related to a symmetry in isospace, it seemed reasonable to look for a symmetry involving both isospin and strangeness. The attempt to find such a symmetry began in 1955, but was not successful until a few years later with the establishment of the *eightfold way*, which will be discussed in Section 4.2.

2.3 Particles Slightly out of Tune

The new quantum number, strangeness, led to such strange behavior of the neutral K mesons that Gell-Mann and Pais did not dare to present it at a conference on nuclear and meson physics in 1954. This behavior has its origin in the fact that there are two neutral strange mesons: the K^0 meson, with strangeness -1, and its antiparticle, the $\bar{K}^0$, with strangeness $+1$. In the weak decay of the K mesons, strangeness is not conserved but only—at least to very good approximation—charge parity; see Section 1.5.4. It is a particular feature of quantum physics that the superposition principle holds: the sum of two states is again a state. Let us consider the superposition of a K^0 meson state with the state of its antiparticle $\bar{K}^0$ as a sum and as a difference:

$$K_S = K^0 + \bar{K}^0 \quad \text{and} \quad K_L = K^0 - \bar{K}^0 .$$

Applying charge conjugation to K^0 yields $\bar{K}^0$ and vice versa. Therefore, the sum K_S remains unchanged on application of charge conjugation, whereas K_L changes its sign. We say that K_S has positive and K_L has negative charge parity. A neutral state formed from two pi mesons always has charge parity $+1$, and therefore only the superposition K_S can decay into two pi mesons; the K_L state cannot. The latter has to decay into at least three pi mesons.

The more kinetic energy available for the decay products, the faster the decay. The kinetic energy available for decay into two pi mesons is proportional to the mass of the K meson minus twice the mass of the pi meson, that is, 217 MeV. In the case of decay into three pi mesons, the energy is only 80 MeV. Therefore, the lifetime of the K_L is 0.09 nanoseconds, whereas that of the K_L is 52 nanoseconds. These different lifetimes lead to the strange concept of particle mixing, about which Pais has written, "The . . . concept of 'particle mixtures' was so unfamiliar that it was thought best not to report it in the Glasgow conference." In reality it is not so complicated. We can solve the above equations involving K_L and K_S for K_0 and obtain

$$K^0 = \tfrac{1}{2}(K_L + K_S),$$

since the $\bar{K}_0$ cancels in the sum. The K_S part decays very rapidly, and after a few tenths of a nanosecond, only the K_L part remains. After a short time, a beam of K_0 mesons has become a mixture (or better, a superposition) of K_0 and $\bar{K}_0$.

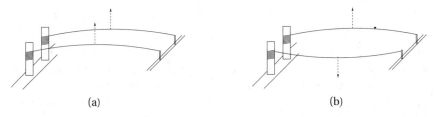

(a)   (b)

Figure 2.5. (a) Two strings with in-phase oscillations and (b) two strings with out-of-phase oscillations in a piano. The in-phase mode is strongly damped (it is analogous to the K_S), while the out-of-phase mode transfers little energy to the soundboard and is therefore weakly damped (it is analogous to the K_L).

This unfamiliar concept has a nice analogue in acoustics. In a piano's middle and upper registers, two or three strings sounding in unison are employed for each note. This makes the sound not only more audible, but also more lively, as we shall see. The two or three strings can either vibrate in phase, as shown in Figure 2.5(a), or out of phase, as shown in Figure 2.5(b). If only one string is struck, it can be viewed as a superposition of the parallel and antiparallel vibrations, with the effect of one of the strings being cancelled by the string oscillating in the opposite phase, as illustrated in Figure 2.6. The strings vibrating in phase transfer much more energy to the soundboard than those out of phase: in the former case the two strings push or pull the soundboard in unison, while in the latter case one pushes while the other pulls. Therefore, the parallel (in-phase) vibration is much more audible and also much more strongly damped than the out-of-phase vibration. We thus observe the following phenomenon: even if only one string is struck (*una corda* on a grand piano), after some time, both strings will vibrate out of phase, and this makes the sound very soft. One oscillating string corresponds to a K^0 meson, while the two out-of-phase strings are analogous to a K_L. The reason for this analogy between particle physics and acoustics is based on the fact that both are described by field theories in which the superposition principle holds.

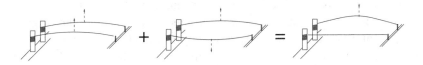

Figure 2.6. The superposition of in-phase and out-of-phase vibrating strings corresponds to one string at rest and one string vibrating.

This example is instructive in yet another aspect: the unison strings of a piano note usually are not exactly in tune; the result is beats, or pulsations in the sound, which make a piano tone more lively. One can interpret the beats as an oscillation between in-phase and out-of-phase vibrations. In the case of K mesons, the detuning of the strings corresponds to a mass difference between the K^0 meson and the $\bar{K}^0$ meson. Even a very small mass difference leads to a beat between K_L and K_S: a beat of 1,000 Hz corresponds to the tiny mass difference of 4×10^{-18} MeV/c^2. This effect makes it possible to determine the mass difference between the K^0 and its antiparticle with extreme precision. For some years now, beats between neutrinos have attracted considerable attention, as we will see in Section 7.1.

2.4   Successes and Failures of Quantum Field Theory

At the end of the 1940s, quantum electrodynamics was brought to a formal completion. Though the infinities could not be removed, physicists learned to live with them, and the predictions of QED were tested with a precision beyond that achieved in any previous theory. There were occasionally experimental hints that the theory was incorrect, but all contests between theory and experiment ended with a clear victory for theory.

But not all scientists were satisfied with the pragmatic way of dealing with the problems of quantum field theory. Dirac, one of the founders of the theory, disliked it all his life and judged the theory to be unaesthetic and incomplete. It is an irony in the history of science that today, even after all the triumphs of quantum field theory in accounting for the electromagnetic, weak, and strong interactions, this opinion is more widespread than it was 20 years ago. We shall return later to this point.

We have already seen in Section 1.4.2 that in many cases, we have to rely on perturbation theory. We begin with a quantized free theory, that is, a theory without interactions, and regard the interactions as a small perturbation. As an example of the results of perturbation theory one may cite the magnetic moment of the electron. If one takes quantum corrections into account, that is, the creation and annihilation of virtual photons and electron–positron pairs, then for the magnetic moment μ, one obtains, from theoretical calculations,

$$\mu_{\text{theory}} = \mu_B \left(1 + 0.5a - 0.32847844400a^2 + 1.181234017a^3 - 1.5098a^4 \right).$$

Here a is essentially the square of the electric charge of the electron, $a = e^2/(4\pi^2\hbar c) = 0.0023228\ldots$. The quantity μ_B is the Bohr magneton, the value of the magnetic moment in the free theory, $\mu_B = \frac{e\hbar}{2m_e}$. This μ_B is the value obtained by Dirac in his theory without any quantum corrections. The second term in parentheses on the right-hand side stems from the quantum correction displayed graphically in Figure 1.16(b). The third term has been known for over 50 years, and the fourth term has now been obtained analytically. For the fifth term, some integrals have been calculated only numerically. With each order of accuracy, the computational cost increases considerably; for example, the a^4 term contains 891 Feynman diagrams, each diagram corresponding to a complicated multidimensional integral. Yet experimental physicists are in no better position than the theorists. In order to test the predictions of theory to this level of accuracy, very refined techniques are necessary. Such techniques were developed and applied by the Nobel laureates I. I. Rabi, P. Kusch, H. G. Dehmelt, and W. Paul. The most precise experiments were performed with a single electron caught in a magnetic trap. The current best experimental and theoretical values are

$$\mu_{\text{experiment}} = 1.001\,159\,652\,18\frac{e\hbar}{2m_e},$$

$$\mu_{\text{theory}} = 1.001\,159\,652\frac{e\hbar}{2m_e}.$$

The agreement between theory and experiment is remarkable.

I have already mentioned that the detection of the pi meson in 1947 was considered an important step toward an understanding of nuclear forces in the framework of field theory. And indeed several remarkable results were achieved. The charge independence of nuclear forces could be explained simply and precisely. The important experimental result that the nucleus of heavy hydrogen, consisting of a proton and a neutron, has the shape of a football and not of a squashed sphere could be explained in the framework of meson field theory, from which one could conclude that the pi meson has spin zero and that the quantum field changes its sign under spatial reflections. One calls particles with this property *pseudoscalars*. A *scalar* particle is also spinless, but its field remains unchanged under spatial reflections. One could also qualitatively explain that the observed value of the magnetic moment of the proton is very different from the result obtained in a free theory. The proton can pass over to a virtual neutron and a pi meson for a short time, and this contributes to the quantum correction to the magnetic moment. A Feynman diagram of the

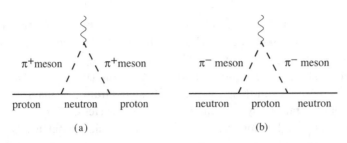

Figure 2.7. Feynman diagrams for the magnetic moment of the nucleons. (a) For the proton. (b) For the neutron. The wavy line represents the electromagnetic field in which the magnetic moment is measured. The positively (negatively) charged pi meson is denoted π^+ (π^-).

quantum corrections is displayed in Figure 2.7(a). The same mechanism explains why even the uncharged neutron can have a magnetic moment. What is decisive here is the transition of the neutron into a virtual proton–pi meson pair; see Figure 2.7(b).

Several researchers calculated the magnetic moment of the proton and neutron. In 1949, S. Borowitz and W. Kohn applied the renormalization scheme just developed by Schwinger (see Section 1.4.2) to the meson field theory. The results were finite. That is, the renormalization procedure was working and the magnetic moments of the neutron and proton had opposite signs, as observed experimentally. However, the numerical results were disappointing. Taking into account only the first term, corresponding to Schwinger's result $a/2$ for the magnetic moment of the electron, one had to take a nucleon–pi meson coupling $g_{N\pi}^2/(4\pi\hbar c)$ of 7 to reproduce the magnetic moment of the neutron, and the value 50 in order to obtain the magnetic moment of the proton. The unsatisfactory agreement was not astonishing. For the magnetic moment of the electron, the first quantum correction, $a/2$, changes the value of the free theory only by 0.1%. The higher powers are correspondingly much smaller. The measured value of the proton's magnetic moment is greater than the value of the free theory by a factor of 2.8. Correspondingly, in that case, a has the value 3.6, which is by no means a small perturbation, and the higher-order terms become more and more important. Therefore, one could not expect that a perturbation theory would lead to quantitatively reliable results. By the same token, it was not astonishing that predictions for the scattering of pi mesons on nucleons did not agree with experiment. One thing, however, was certain: the nuclear forces had to be strong, and slowly the name "strong interactions" for nuclear interactions gained currency. Now opinion began to diverge. One group considered quantum field theory to be completely useless for strong interactions. Others believed that quantum field theory offered the only possibility for explaining the properties of elementary particles, and they tried first to extract every possible conclusion from field theory. This rather conservative attitude, which more than 20 years later proved to be enormously successful, concentrated at first more on general than on particular results. Symmetry considerations played a decisive role in this approach, as we will see in Section 2.5.

Another important development was the search for general results in an axiomatic field theory. Since a precise formulation would far exceed the scope of this book I will try to express the Wightman axioms of quantum field theory in everyday language. The following are the general postulates for a local, relativistically invariant quantum field theory:

W1 (Covariance). In all uniformly moving frames of reference, the field equations have the same form.

W2 (States and Observables). All states can be generated from the vacuum state by application of field operators; the vacuum is the same in all frames of reference. The observables can be expressed through field operators. This is the program of quantization.

W3 (Causality or Locality). Field operators acting on space-time points that cannot be connected by light signals are completely independent.

W4 (Uniqueness of the Vacuum). The above-mentioned vacuum state is the only one that does not change over time.

These postulates were used to derive a whole series of far-reaching theorems. One of those is the so-called *CPT theorem*; it was proved independently in 1955 by J. Schwinger, G. Lüders, and W. Pauli. It states that every quantum field theory is invariant under the simultaneous application of charge conjugation (C), space reflection or parity (P), and time reflection (T); thus the name of the theorem. This means that in a world in which particles are replaced by antiparticles, and left-handed objects by right-handed objects, and where time runs backward, the same laws of nature hold as in our world. This sounds rather esoteric, but a direct consequence of the theorem is that the masses of particles and antiparticles are exactly equal. The mass difference between particle and antiparticle has been measured with great precision for the electron and the K^0 meson. The numbers show how precisely the theorem is verified:

$$|m_{e^+} - m_{e^-}|/(m_{e^+} + m_{e^-}) \leq 4 \cdot 10^{-9},$$
$$|m_{K^0} - m_{\bar{K}^0}|/(m_{K^0} + m_{\bar{K}^0}) \leq 10^{-18}.$$

The extreme precision of the K^0 meson calculations is possible only on account of the beat phenomena discussed in the previous section. There are few statements in the exact sciences that have been so precisely verified as those following from the CPT theorem.

Another important theorem that was proved in axiomatic field theory concerns the relation between spin and statistics. A first version of this theorem for theories without interaction was derived by Fierz and Pauli in 1938. The general proof is due to Lüders and B. Zumino as well as to N. Bourgoyne (1958). It states, in part, that particles with half-integer spin obey Fermi–Dirac statistics and therefore satisfy the Pauli exclusion principle. According to this principle, in a state of many particles, no two spin-$\frac{1}{2}$ particles can have the same quantum numbers. This is the reason for the shell structure of electrons in the chemical elements and thus for chemical diversity. Particles with integer spin do not obey the exclusion principle. On the contrary, here a multiparticle state exhibits the tendency that all particles have the same quantum numbers, namely, those that lead to lowest energy. This leads to what is known as Bose–Einstein condensation. While the effects of the exclusion principle are relevant for effects that occur in our everyday lives, Bose–Einstein condensation is a much more exotic effect, occurring only at very low temperatures. The 2001 Nobel Prize in physics was awarded for its first direct observation. Because of their different behavior, particles with half-integer spin are called *fermions*; those with integer spin are called *bosons*. The proofs of these very precisely tested and far-reaching theorems are among the greatest successes of axiomatic quantum field theory.

2.5   The Beginnings of a New Spectroscopy

We now take a step backward in time, to the early 1950s, the time of the first synchrocyclotrons. We already have seen that with the discovery of the neutral pi meson, accelerators had made a first important contribution to particle physics. They turned out to be an indispensable aid in a new branch of research, which can be called the spectroscopy of strongly interacting particles; it led to completely new insights into the fundamental laws of nature.

Lawrence and his group in Berkeley were interested principally in the development of accelerators. In Chicago, where a synchrocyclotron was built in 1951, Enrico Fermi was strongly involved, but he was less interested in the development of accelerators than in performing experiments with them. We were introduced to him in Section 1.4.2 as an eminent theoretician, but he was also one of the most important experimentalists. The experiments that he and his group performed in Chicago largely determined the course of further developments in experimental physics. Protons in the Chicago synchrocyclotron were accelerated to an energy of 450 MeV and directed at an internal target, a block of beryllium; the result was a copious production of pi mesons. At high energies, most mesons are produced in the forward direction; that is, they travel in the same direction as the accelerated protons. The positive pi mesons produced in the internal target are deflected inward by the magnetic field of the accelerator, while the negatively charged pi mesons are deflected outward and thus can be extracted easily. Therefore, the first results obtained with negative pi mesons were more precise than those obtained with positive pi mesons. The magnetic field of the accelerator selected the pi mesons according to their energy. A selected beam with a definite energy was directed at a target of liquid hydrogen, making possible the investigation of interactions between pi mesons and hydrogen nuclei, that is, protons; the interaction with hydrogen electrons is negligibly small. The setup of the experiment is shown very schematically in Figure 2.8.

In the initial experiments, the relative number of pi mesons absorbed by hydrogen was measured. This allowed the researchers to determine the *total cross section*, that is, the area that a pi meson effectively "sees" if it happens

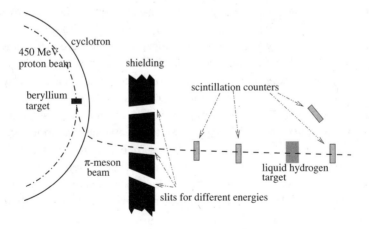

Figure 2.8. Schematic diagram of the scattering experiment performed by Fermi and collaborators at Chicago.

Figure 2.9. The trolley car constructed by Fermi for transporting the beryllium target to the optimal position within the accelerator.

upon a proton. In much more complicated experiments, the *differential cross section* was also measured. This magnitude is related to the probability with which a pi meson is deflected in a certain direction. Fermi actively participated in preparing the experiments. He was in charge of preparing the targets. One weekend, he built a small trolley car on which the internal beryllium target could be moved into optimal position for the desired energy of the pi mesons. Fermi's trolley car is shown in Figure 2.9.

The detectors necessary for such scattering experiments had been invented by H. Kallmann. They were a further development of the simple scintillation counters used earlier by Rutherford and Geiger, but they were made of transparent plastic and therefore had a much larger useful volume. The light flashes induced by the ionizing particles were of course no longer registered by eye, but amplified and registered electronically. The results of the scattering experiments were exciting. First, the simpler experiments with negative pi mesons were performed; the cross section was large, and at an energy of about 140 MeV reached the value expected from a particle the size of the proton. With the scattering of positive pi mesons, on the other hand, the experimenters were caught by surprise: the cross section far exceeded the expected value. At an energy of 120 MeV it had twice the value of that for the negative pi mesons, and there was no apparent end to the increase.

Fermi and his collaborators might have been surprised, but they were not totally unprepared. That very day, they had received a paper by K. A. Brueckner containing a prediction for the total cross section on the basis of earlier exper-

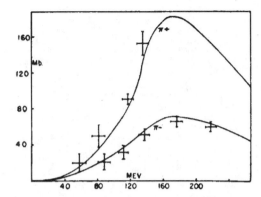

Figure 2.10. Brueckner's predictions for the cross section of positive and negative pi mesons (solid lines) and the experimental results of Fermi and his collaborators. Brueckner had assumed the existence of a resonance state with isospin $\frac{3}{2}$; it follows from isospin symmetry that the cross section for the scattering of positive pi mesons (π^+) is three times that of negative pi mesons (π^-).

iments. This prediction coincided quite well with the results of the new experiments. Figure 2.10 shows Brueckner's predictions together with the results of the Chicago group.

This brings us back to isospin. We saw in Section 1.5.3, Figure 1.23, how to group the four pairs consisting of protons and neutrons, namely $\{p, p\}$, $\{n, p\}$, $\{p, n\}$, and $\{n, n\}$, into a triplet with isospin 1 and a singlet with isospin 0. In the same way, one can group together the six pairs that can be formed from the isodoublet of the nucleons and the isotriplet of the pi mesons into a quadruplet with isospin $\frac{3}{2}$ and a doublet with isospin $\frac{1}{2}$. Brueckner had assumed the existence of a state with isospin $\frac{3}{2}$ and spin $\frac{3}{2}$ with a mass of $1,250\,\mathrm{MeV}/c^2$.

Using group theory, one can easily calculate that the contribution of such a state for the scattering of positive pi mesons on protons is three times that for negative pi mesons. If the incident pi mesons have just the right energy to excite this resonance state, namely 215 MeV, a resonance phenomenon takes place, and the cross section is no longer determined by the geometric size of the proton, but by the quantum-mechanical wavelength of the incident pi mesons, called the *de Broglie wavelength*. Figure 2.11 shows a page of Fermi's notebook in which he uses the isospin formalism. It later emerged that there was not only the one resonance with isospin $\frac{3}{2}$, the so-called isobar or delta resonance. Accessible energies grew higher and higher, detectors became more and more

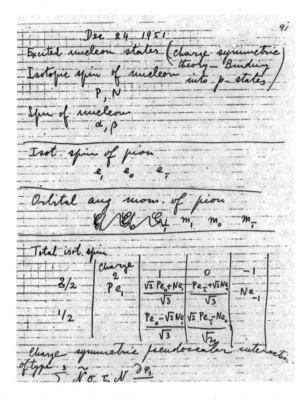

Figure 2.11. A page from Fermi's notebook; the ratio of cross sections for positive and negative pi mesons is calculated using isospin symmetry.

refined, and several hundreds of these resonance states were found, not only for nucleons, but also for mesons. As many resonance states as there are spectral lines in atomic physics were discovered, and the terms "particle spectrum" and "particle spectroscopy" were coined. We will discuss this more in the following sections.

2.6   Producing More and Seeing Better

The big discoveries of the artificial elementary particles—the positron, the muon, the charged pi mesons, and the V particles—began with natural processes, triggered by cosmic rays. The detection methods were old, having been

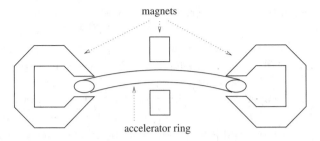

Figure 2.12. Design principle for a synchrotron. The C-shaped magnets are arranged around the acceleration vessel.

developed at the end of the nineteenth century. The first particle to be produced artificially, that is, by high-energy particles accelerated in a manmade accelerator, was the neutral pi meson. But the scattering experiments discussed in the previous section were possible only because a sufficiently intense beam of pi mesons could be produced with the help of an accelerator. In Figure 2.10 one can see that the Chicago experiments had to stop at an especially interesting energy at which the cross section was bound to decrease again—if it was really a resonance phenomenon. It was therefore evident that essential progress in particle physics would be reached only through accelerators with higher energies. The production of V particles also demanded higher energies. However, with the design principle of the synchrocyclotron having soon reached its limits, it became a difficult task to build accelerators with higher energies. The diameter of the magnet would have to be as large as the trajectory of the particles on reaching their final energy, and a big magnet is very expensive. However, the inner part of the magnet is used only to bring the particles to an energy corresponding to the final beam trajectory, as can be seen from the schematic in Figure 1.28. Thus the obvious next step was to omit the inner part of the cyclotron. Preaccelerated particles were injected into a doughnut-shaped vacuum vessel surrounded by magnets, where the particles were accelerated further. This is shown schematically in Figure 2.12. In the old synchrocyclotron, the magnetic field was constant and the radius of a particle's trajectory increased with increasing energy. In the newly developed synchrotron, in contrast, the radius of the trajectory is constant and the magnetic field and acceleration frequency are adapted to the energy of the particles.

Here again, the principle of phase stability was helpful, and it is therefore no wonder that one of the inventors of this principle, the British physicist McMillan, constructed the first accelerator of this type. It was completed in 1949 and it could accelerate electrons to an energy of 320 MeV. The first huge synchrotron was already a national project: the Cosmotron of the Brookhaven National Laboratory (BNL), on Long Island, New York. It was approved in 1948 and completed in 1953. Its diameter was 22.5 meters. The magnetic field was generated by 288 C-shaped magnets, and it could accelerate protons to an energy of 3.2 GeV, that is, 3,200 MeV. It was in use until 1966 and was scrapped in 1969. The next big machines of this type were constructed in Berkeley, California in 1954 (6 GeV) and in Dubna, Russia in 1957 (10 GeV).

Soon, however, these accelerators reached their limits as well. The trajectories of the particles in the acceleration vessel were only weakly focused and therefore rather unstable. Oscillations around the circular trajectories grew, with the result that the particles hit the walls of the vessel and were thus lost to the acceleration process. In 1952, R. Courant, M. S. Livingston, and H. Snyder discovered the *strong focusing principle* by which the trajectories could be stabilized. This is achieved by an ingenious arrangement of the magnets and their inhomogeneities. One magnet focuses in the horizontal direction, the next magnet in the vertical. These accelerators are therefore called strong focusing or alternating gradient (AG) machines. By avoiding the vibrations around the circular trajectories, the cross section of the acceleration vessel and thus the magnets can be made much smaller than in the old weakly focusing machines.

The first accelerator designed according to this new principle was built at Cornell University, in Ithaca, New York; it was able to accelerate electrons to an energy of 1 GeV. The next big accelerator was built at the Brookhaven National Laboratory around 1960. It had a final energy of 33 GeV; the cross section of the accelerator vessel was 17.5 cm × 7.5 cm, and the weight of the magnets was only four thousand metric tons. For the 3.2-GeV Cosmotron that was completed in 1953, the cross section of the vessel was 65 cm × 15 cm, while the mass of the magnets was two thousand tons. If the new machine had been constructed according to the old weak focusing principle, it would have required about two hundred thousand tons of iron. The first big accelerator in Europe became feasible through a multinational effort. In 1952, eleven European nations founded the European Organization for Nuclear Research (CERN), which soon became one of the leading research centers in particle physics. Abraham Pais writes in his beautiful book *Inward Bound,* "Now (1987) ... CERN ranks among the very few leading centers in the field, has produced the highest effective energy reached thus far in a laboratory, and stands as the most successful example of what Europe can achieve by pooling its resources."

In addition to progress in accelerator physics, detection methods were also greatly improved. We saw in the last section that large scintillation counters with electronic amplification and registration of the signals were invented just in time. But the most significant advances made in particle physics in early modern times were due to a refinement of the cloud chamber, which led to the invention of the bubble chamber in 1952 by D. A. Glaser. The bubble chamber contains an overheated liquid, and evaporation germs are generated along the trajectory of an ionizing particle, giving rise to small bubbles. These mark the trajectory just as water droplets did in the cloud chamber. Since the density of a liquid is great compared to that of a gas, the bubble chamber has two decisive advantages over the cloud chamber: (1) many more reactions occur and (2) the tracks are shorter and therefore a full reaction chain can frequently be observed. Bubble chambers thus unite the large volume of a cloud chamber with the high density of a photographic emulsion.

The first bubble chambers were filled with a liquid with a high boiling point, such as propane. Later, large chambers filled with liquid hydrogen at a temperature of $-250°$ C were built. These were very expensive and dangerous devices, but were ideally suited for studying elementary reactions on protons. As in the case of the cloud chamber, the tracks are photographed with two cameras to enable the spatial reconstruction of the particles' trajectories. Scanning the photographs was cumbersome, just as in the case of photographic emulsions. In spite of great progress in the automation of the photographic analysis, bubble chambers were eventually substantially replaced by a new type of counter, the wire chamber, for which digitization and automatic data processing were much easier. We will return to this topic in Section 5.6.

2.7   More and More New Particles

New accelerators and more refined detection devices contributed to the consolidation of quantum field theory. In 1955, a group formed around O. Chamberlain and E. Segrè found the first antiparticle with baryon number −1: the antiproton. It was produced at the Bevatron, a 6-GeV accelerator in Berkeley. In addition, more and more resonances such as the delta resonance discussed in Section 2.5 were detected. The energy (mass) of these resonances is not well defined because of their short lifetime. The energy uncertainty (width) of these resonances is typically in the range of ten to a few hundred MeV, corresponding to a lifetime in the range 10^{-22} to 10^{-23} seconds, according to the energy–

time uncertainty principle. As can be seen from Table 2.3, there was a flood of newly detected resonances. Detection and determination of their properties were possible only with the help of the bubble chambers, especially those filled with liquid hydrogen. Here the traces of all charged particles are visible, and the momentum of the particle can be calculated from the curvature in the magnetic field. It was L. Alvarez, of the University of California, Berkeley, who contributed greatly to the further development of bubble chambers and the refinement of detection methods.

The first mesonic resonances, that is, states with baryon number zero, were detected in the bubble chamber. Consider the reaction in which a negative pi meson strikes a proton and thereby creates an additional positive pi meson. Because of charge conservation, the proton must be transformed into a neutron: only then does the total charge remain the same as that of the initial state, namely zero. The shorthand notation for such a reaction is, in analogy to chemical formulas, $\pi^- + p \to \pi^- \pi^+ + n$. From the momenta of the two pi

year	name	mass MeV/c^2	width/ MeV	B	S
ca. 1951	Λ	1115.7	–	1	−1
ca. 1951	K	ca. 495	–	0	±1
ca. 1952	Δ	1232	120	1	0
1952	Ξ^-	1321	–	1	−2
1953	$\Sigma^\pm$	ca. 1190	–	1	−1
1959	Ξ^0	1315	–	1	−2
1960	Y^*	1385	36	1	−1
1961	K^*	892	51	0	±1
	ρ	770	150	0	0
	ω	782	8	0	0
	η	547	0.001	0	0
1962	Ξ^*	1530	9	1	−2
	f^0	difficult		0	0
	ϕ	1020	4.4	0	0
1964	Ω^-	1672	–	1	−3
	η'	958	0.2	0	0
⋮	⋮	⋮	⋮	⋮	⋮

Table 2.3. The dates of discovery and basic properties of elementary particles. For resonances, the width is indicated; if the width is not given, the particles cannot decay through strong interactions and their lifetime is in the range of nanoseconds. B is the baryon number, S the strangeness.

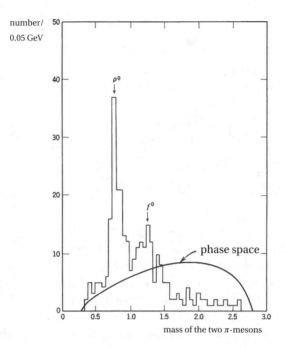

Figure 2.13. The mass of the system consisting of a positive and a negative pi meson. The system was produced by scattering of negative pi mesons on protons. The mass of the system can be varied by changing the energy of the incident pi meson beam. The resonance at 770 MeV/c^2 is clearly visible. The solid line marked "phase space" shows how the system would behave if there were no resonance.

mesons after the reaction one can determine the mass of the two-pi-meson state, and one finds a clear resonance behavior; see Figure 2.13. The baryon number of the resonance is zero. From the distribution of the two pi mesons one can conclude that the spin of the resonance is 1 and the internal parity -1. Such a state is called a *vector meson*. In Figure 2.13, a second resonance structure is visible, but its interpretation is more difficult. In addition to this neutral meson with resonance at 770 MeV/c^2, in other reactions a positively and a negatively charged meson with approximately the same mass were found. This led to the conclusion that the isospin is 1. Such a resonance is called a rho meson (ρ-meson). One can represent the resonance production in a diagram as in Figure 2.14, where the zigzag line corresponds to a rho meson. We will return frequently to these resonances and their role in the system of elementary particles.

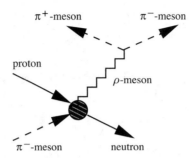

Figure 2.14. Diagram for the production of a rho meson (ρ-meson).

Resonances were not a new phenomenon, even in microscopic physics. The famous experiment studying the scattering of electrons on mercury atoms, performed in 1913 by James Franck and Gustav Hertz, was a crucial experiment in quantum mechanics. Franck and Hertz found that the cross section for electron–mercury scattering was especially large if the energy of the electrons was just sufficient to produce the known state of the mercury atom with an excitation energy of 4.9 eV. Every neon light uses such resonance phenomena.

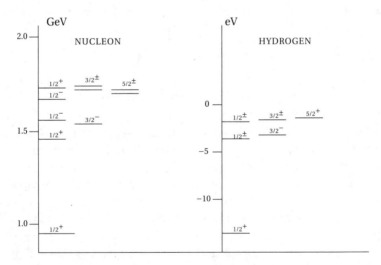

Figure 2.15. Resonances of the nucleon (baryon number 1, strangeness 0, isospin $\frac{1}{2}$) and excited states of the hydrogen atom. For the different states, spin and internal parity are indicated. For example, $5/2^+$ means spin $\frac{5}{2}$, parity $+1$.

The many resonances since detected in particle physics are indeed strongly reminiscent of the lines of atomic spectra, as can be seen from Figure 2.15. On the left, the resonances of the nucleon—baryons with isospin $\frac{1}{2}$ and strangeness 0—are displayed. The fractional number indicates the spin of the resonance, while the upper index is the internal parity. The excited states of the hydrogen atom, which can be calculated with the Dirac equation and have been precisely measured, are displayed to the right. Here spin and parity are indicated as well. The structure is similar, but the scale is different by a factor of nearly a billion; the excitation energies are on the order of GeV for the particles and eV for the atoms.

The existence of the many resonances gave a hint that the large number of newly detected particles were composites. The stable particles in such a scheme should correspond to the atoms; the resonances, to excited states. This is indeed the present consensus. Ideas along these lines appeared as early as 1949, and we shall return to this topic later on. There is, however, a tremendous difference between atomic physics and particle physics. In atomic physics, the constituent parts of the atoms were well known: the nucleus and the electrons of the surrounding cloud. The resonances could be explained as excitations of the cloud and are well described in the framework of quantum mechanics.

What is it, then, that is excited in the case of nucleons? Is the proton a bound ground state of an original proton and a pi meson cloud? Are the resonances excitations of the pi meson cloud? Are the mesons themselves bound states of nucleons and antinucleons? To all these questions there were no satisfying answers. The difference in the excitation energies mentioned above prevented the known theoretical tools of atomic physics from being applied. The typical excitation energy of an atom is a few electron volts, much smaller than the rest energy of the electron at $500,000$ eV. Particle production therefore plays only a minor role. In fact, the quantum corrections in atomic physics are very small, similar to the corrections to the magnetic moment of the electron, discussed in Section 2.4. Nevertheless, quantum corrections can lead to new effects in atomic physics as well. W. E. Lamb discovered a splitting of energy levels of the hydrogen atom that could not be explained by the Dirac equation but is rather a consequence of quantum field theory.

The situation is completely different for elementary particles. The excitation energies are in general larger than the rest energy of the lightest particle, the pi meson (around 140 MeV). Therefore, one expects particle annihilation and creation to play an important role. As a consequence, the methods of perturbation theory, which were so successful in QED, cannot work with comparable precision, as has been discussed already in Section 2.4.

Nonperturbative models were constructed as a reaction to this shortcoming. In these models, more-or-less stringent approximations were made in order to make them solvable. A model of T. D. Lee and one of G. Chew and F. E. Low yielded some fundamental qualitative insights, but no satisfying quantitative results. Many physicists came to believe that quantum field theory should not be applied to the strong interactions of elementary particles and that an entirely new approach was needed. We will discuss these new approaches in the next chapter, but before doing so, we return to the weak interactions.

2.8   The Surprises of the Weak Interaction

The theory of weak interactions, initiated by Enrico Fermi in 1933, was one of the first successful applications of quantized field theory. In spite of its successes, it gradually became clear that it could not be a complete theory. In calculating higher-order corrections in perturbation theory, difficulties were encountered that could not be tamed by the renormalization formalism that had meanwhile been developed in QED. This meant that the theory was not renormalizable; calculations had to be confined to the lowest-order perturbation theory. But there were many new experimental facts that could be treated in this lowest order, and theoreticians did not worry too much about the fundamental incompleteness and were fully occupied in extending the lowest-order perturbation theory; very important contributions were made by G. Gamow, E. Teller, E. Majorana, and B. Pontecorvo.

We saw in Section 2.2 that the weak decays of the strange V particles led to the introduction of a new charge-like property of elementary particles, the strangeness S. It is conserved in strong and electromagnetic interactions, but violated in weak interactions. But there was soon to be a big surprise in the weak interactions, caused by the *theta–tau puzzle*. Before the classification of K meson decays was clarified, two kinds of strange charged mesons were known: the tau (τ) and the theta (θ). The tau meson (τ-meson) decayed into three pi mesons. Figure 2.3 shows the traces of the K meson and its decay products. To analyze these decays, R. H. Dalitz developed a technique now called the Dalitz plot. It played an important role in the analysis of particle decays. Dalitz plotted the domain allowed by energy and momentum conservation in such a way that equal areas of the plot corresponded to equal areas of phase space. This means that if the distribution is not uniform, there is a dynamical reason for it. The more entries there are in the plot, the better, since a single particle cannot be equally distributed. By clever use of symmetries, Dalitz was able to increase the density of points in the plot, and from 13 decays he concluded that the tau meson has spin 0. The inner parity of the tau meson was thus fixed to be −1, namely, the product of the internal parities of the three pi mesons occurring in the decay. By 1955, 55 decays had been analyzed, 11 of which were from accelerators. One year later, 600 decays were observed, the vast majority coming from accelerators. With these large numbers, the earlier results were established beyond any doubt: the tau meson had internal parity −1.

The other charged strange meson, called the theta meson (θ-meson), had within experimental error the same mass and the same lifetime. In spite of this congruence, however, it was thought that they must be different particles, since the theta meson decayed into two pi mesons and thus had inner parity +1, in contrast to the internal parity of −1 for the tau meson. But the agreement between the masses and, especially, the lifetimes was a tantalizing riddle. After seeing Dalitz's report, Oppenheimer remarked, "The tau meson will have either domestic or foreign complications. It will not be simple on both fronts." Shortly thereafter, Abraham Pais, noted, "Be it recorded here that on the train from Rochester to New York, Professor Yang and the writer each bet Professor Wheeler one dollar that the theta- and the tau-meson were distinct particles; and that Professor Wheeler has since collected two dollars."

The solution to the riddle came in 1956 through the work of T. D. Lee and C. N. Yang. They showed that there was no experimental evidence for symmetry of the laws of weak interactions under space reflection. But if the interaction is not invariant under space reflections, then the internal parity of the decay products need not be the same as that of the decaying state, and therefore the theta and tau meson could very well be the same particle. Lee and Yang also proposed an experiment to prove *parity violation* in weak interactions directly.

The first results of such an experiment were published in 1957 by Madame C.-S. Wu and her collaborators. At low temperatures and with high magnetic fields, they aligned the nuclear spins of the radioactive cobalt isotope (Co^{60}) predominantly in one direction. They observed that more electrons were emitted in the direction of the nuclear spin than in the opposite direction. Under space reflection, the direction of flight changes sign, but not the spin. The direction is a vector, while the spin is an axial vector. A space-reflected world would therefore be different from ours, since in such a world, more electrons would fly in the direction opposite to the nuclear spin. Today in particle physics, one has become accustomed to such ideas—as one has to many other ideas—but at the time, this was a true sensation. The nuclear physicist and later Nobel laureate J. H. D. Jensen said that while visiting the laboratory of Madame Wu, he teased her to the effect that American physicists have too much money, since they are developing a complicated experiment whose outcome was clear. Of course, he thought that parity was conserved and that the outcome of the experiment would be negative. Parity violation was soon confirmed by many other experiments. In spite of the surprise, the shock was limited. Soon the theory of weak interactions gained elegance, and a new symmetry appeared. But for that, I have to return to group representations, which were discussed briefly in Section 1.5.1. I will only report the results and will not give the derivations.

2.8.1   Digression: Right- and Left-Handed Particles

We have seen in Section 1.5 that a group is determined by its generators. The commutators of the generators set the group structure to a large extent, but not completely. In Section 1.5.2, we saw that the same generators with the same commutation relations can generate both the rotation group and the group SU(2). This led us to the miracle of spin: the spin group SU(2) is richer than the rotation group. A physical consequence is that the rotation group admits only integer angular momenta, but the group SU(2) also admits half-integer spins.

We now generalize rotations in three-dimensional space to the space-time transformations that obey the principles of special relativity. Under these transformations, not only angles and lengths remain invariant, but also the velocity of light in a vacuum. This group is called the *Lorentz group*, after the Dutch physicist Hendrik A. Lorentz. Its fundamental representation is four-dimensional, because of the three space dimensions and one time dimension. The four-dimensional matrices are called Lorentz transformations. With this group, the miracle of spin repeats itself. Once again, there is a more general group, which can be represented by 2×2 matrices, from which all representations of the Lorentz group can be constructed. This was recognized in 1929 by the mathematicians B. van der Waerden and Hermann Weyl. Weyl in particular pointed out that for the doublets transforming under that two-dimensional representation one can construct a relativistically invariant wave equation that is simpler than the Dirac equation. These doublets are now called *Weyl spinors*.

In 1933, Wolfgang Pauli wrote a famous review entitled "The General Principles of Quantum Mechanics," known to experts as the "New Testament." There he referred to the Weyl spinors and their wave equation. Pauli wrote, "These wave equations are, however, not invariant under reflections (exchange of left and right) and therefore not applicable to physical reality. The absence of invariance of the wave equation manifests itself in a peculiar coupling of the direction of the spin-angular momentum and the current." The argument seemed to be valid at the time, but after it was realized that weak interactions are not invariant under space reflections, Pauli's curse was transformed into a blessing. Today, the description in terms of Weyl spinors is considered to be more fundamental than that involving Dirac spinors. Indeed, the latter can be constructed from Weyl spinors.

There exist two completely independent kinds of Weyl spinors; mathematically speaking, there are two unitary inequivalent representations. Both describe spin-$\frac{1}{2}$ particles. For one kind, the spin is parallel to the direction of movement; these are the *right-handed spinors*. For the other kind, the spin is antiparallel; these are the *left-handed spinors* (see Figure 2.16). This property is called handedness, or *chirality*. Particles and antiparticles have opposite chirality: if a particle is described by a right-handed spinor, then its antiparticle is described by a left-handed one.

Clearly, a particle with definite chirality must be massless: if a particle has a mass different from zero, its velocity is always less than that of light, as a consequence of which it can be overtaken by an observer. If the observer is faster

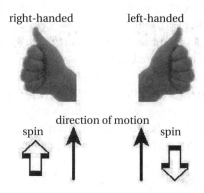

Figure 2.16. Handedness (chirality) of particles. A particle whose spin is parallel to the direction of movement is called right-handed. If spin and direction of movement point in opposite directions, it is called left-handed. The thumb indicates the direction of movement, while the other fingers indicate the rotational movement of the spinning particle.

then the observed particle, the particle moves backward in the observer's frame of reference, exactly like the situation of one automobile overtaking another. Seen from the passing lane, a right-handed particle becomes a left-handed one, since the direction of movement has changed, but not the spin. A massive particle can therefore not be described by a Weyl spinor. Therefore, Dirac had to introduce spinors with four components. The four-component Dirac spinors are composed of two Weyl spinors—a right-handed one and a left-handed one. Transformations that transform right- and left-handed particles separately are called *chiral* transformations, and the corresponding symmetry is *chiral symmetry*. We will return to this topic in Chapter 7.

2.8.2   And Now Back to Weak Interactions

For the moment, I will ignore a discovery dating from the beginning of the new millennium, namely, that neutrinos most probably have a very small mass. (I will return to the question of neutrino masses in Section 7.1.) Let us therefore begin with massless neutrinos, which are a very good approximation. Massless neutrinos are evidently excellent candidates for description via Weyl spinors, and thereby not only is parity violation unavoidable, but also the form of the interaction is determined, the so-called V-A interaction. This shorthand notation means that the weak interaction is described by two different currents on an equal footing. One is an ordinary vector, V, which changes sign under space reflections; the other is an axial vector, A, which does not change direction under space reflections. Since two quantities that behave differently under space reflections occur together with equal weight, parity violation is said to be maximal. After some initial experimental confusion, this type of interaction was indeed confirmed by experimental results.

The part of the current that behaves like a vector is not influenced by strong interactions; it is called universal. But for the axial vector a partial universality holds as well. This sounds rather vague, but it has a well-defined meaning: in the limit of massless pi mesons, the axial part of the current is universal, and deviation from universality can be expressed in terms of the relatively small pi meson mass.

A further experimental result in weak interactions had been generally expected, but it was nevertheless extremely important for the further development of physics. This was the direct observation of neutrinos, the particle postulated by Pauli in 1931. It was evident that detecting neutrinos directly would be an extremely difficult task. In 1934, Hans Bethe and Rudolf Peierls had estimated the cross section for the reaction in which an antineutrino interacts with a proton and is transformed into a neutron and a positron. They found that the cross section is extremely small. They estimated that a neutrino has to pass through more than 1,000 light years of dense matter before an interaction takes place with reasonable probability. Therefore, many, many neutrinos are needed in order to detect a reaction in a realistic counter—with a volume of, say, one cubic meter—in one year.

Frederick Reines first thought of using neutrinos created in a nuclear explosion, but later gave up on that idea and switched to a nuclear reactor. The fission products of nuclear reactions have too few protons to achieve stability. Therefore, neutrons in the nuclei decay into protons and emit electrons and antineutrinos. As a result, a nuclear reactor is a very efficient steady source of antineutrinos. In order to detect antineutrinos, Reines and Clyde Cowan used inverse beta decay, a reaction in which an incoming antineutrino reacts with a proton and creates a neutron and a positron. This reaction is described by the graph in Figure 1.13, but the direction of the arrows must be reversed. The counters for detecting this reaction were screened by many hundreds of tons of lead and by a mixture of boron and paraffin, in order to avoid contamination through events from cosmic rays. Alas, these efforts were insufficient for reliably isolating the extremely rare reaction that they hoped to detect.

Finally an ingenious method of delayed coincidence made it possible to identify the neutrino unequivocally. After the reaction of the antineutrino with the proton in a container of water, the produced positron annihilates together with an electron into two photons, each with an energy of 0.511 MeV. These two photons are detected in coincidence. The produced neutron is slowed down in the water and is absorbed several microseconds later by dissolved cadmium. This leads to a nuclear reaction in which again photons are produced. The photons of this reaction are also detected in coincidence, but this reaction has to occur later than the detection of the annihilation photons. In this manner the neutrino reaction was identified beyond any doubt. Three events per hour were observed. The cross section calculated from this occurrence corresponded well with the estimate of Bethe and Peierls. Fermi was very interested in the experiment, but he did not live long enough to see the direct proof of the existence of neutrinos. But Reines and Cowan had the pleasure of sending the following telegram to Pauli on June 14, 1956: "We are happy to inform you that we have definitely detected neutrinos . . . by observing inverse beta decay. Observed cross section agrees well with expected: six times ten to the minus forty four square centimeters. Frederic Reines and Clyde Cowan."

With their pioneering experiment, Reines, Cowan, and their collaborators had begun a new branch of experimental particle physics, that of neutrino scattering, giving fresh impetus to the theory of weak interactions and particle physics in general.

The refinement of Fermi's theory through the introduction of Weyl spinors makes parity violation very natural. On the other hand, as discussed in Section 2.4, there is the CPT theorem, which states that each local field theory is invariant under the transformation $\mathbf{C} \cdot \mathbf{P} \cdot \mathbf{T}$, that is, the successive application of time reflection $\mathbf{T}$, space reflection $\mathbf{P}$, and charge conjugation $\mathbf{C}$. After parity violation had been firmly established, it was nevertheless generally assumed that weak interactions were invariant under $\mathbf{C} \cdot \mathbf{P}$, a successive application of space reflection and charge conjugation. However, some time later it was found that this combined symmetry is also weakly violated. I shall move somewhat ahead in the chronological order of events and briefly describe this next surprise of the weak interactions.

In Section 2.2 we introduced the states K_S and K_L as superpositions of the neutral K meson and its antiparticle. Moreover, we saw that the state K_L cannot decay into two pi mesons, since it has charge parity -1, whereas a two-pi-meson state has charge parity $+1$. This holds generally if the interaction is invariant under the above-mentioned combined application of space reflection and charge conjugation, $\mathbf{C} \cdot \mathbf{P}$.

Then in 1964, V. L. Fitch, J. W. Cronin, and their collaborators published a paper in which they showed that the long-lived component K_L can also decay into two pi mesons. The fraction of these decays was very small, but beyond doubt. Thus in weak interactions, the combined transformation $\mathbf{C} \cdot \mathbf{P}$ is also not a symmetry transformation; one says that CP is violated. This violation, however, is not maximal, as in the case of space reflections, but very small. Although there is much more disposable kinetic energy for the two-particle decay than for the three-particle decay, the fraction of K_L states decaying into two pi mesons is only 0.3% .

Even today, 40 years after its discovery, CP violation is still a mystery, although it can be parameterized in the present standard model of particle physics (see Chapter 6, especially Section 6.10). A full solution to this riddle would be a giant step forward in our understanding of particle physics. Violation of CP symmetry also implies violation of T, time reflection symmetry, since the rigorous and well-tested CPT symmetry holds.

3

Up by Their Own Bootstraps

A certain disillusionment followed the euphoric mood induced by the discovery of the predicted meson and the successes of quantum electrodynamics. The number of newly discovered particles continued to increase, and in strong interactions, quantum field theory was less predictive than had been hoped. A new way forward seemed necessary.

3.1 S-Matrix Theory

In the last chapter we saw that in the middle of the twentieth century a mood of optimism prevailed. Refined experimental techniques and theoretical methods seemed to confirm the concepts developed in the first half of the century. The amazing discoveries of quantum field theory, such as the existence of antimatter and the possibility of particle creation, had been confimed experimentally. For a time, it seemed that a final theory of elementary particles would soon be established. But as we have seen, not all the dreams were realized. Experimental results clearly showed that nature was not being cooperative—nature was not as simple as the physicists had imagined it to be. "Nobody had ordered" strange particles, but nevertheless, there they were. Theoretical resources were also more limited than had originally been expected.

Application of quantum field theory to strong interactions gave several valuable qualitative insights, but unfortunately no quantitatively satisfactory results. The Fermi theory of weak interactions was very successful, but only if restricted to first-order perturbation theory. Even in QED, where the very precise predictions of the theory had been experimentally confirmed, the situation was not entirely satisfactory. Landau and Pomeranchuk went so far as to call this theory logically incomplete. Their reasons for this were both justified and unjustified. It is an irony of history that the considerations that led to these bleak prospects for quantum field theory were decisive for its splendid resurrection in the 1970s. We will return to this topic in Section 6.5.

These fundamental arguments, but even more the impossibility of moving beyond perturbation theory and making quantitative statements, led many particle physicists to doubt the utility of quantum field theory.

Nevertheless, physicists cannot live without theory. Some of them became more modest and fell back on an important paper by Heisenberg, published in 1943. Titled "Observable Quantities in the Theory of Elementary Particles," it had been written before the development of renormalization theory (see Section 1.4.2), and Heisenberg thought that the difficulties of quantum field theory could be solved only if "the future theory contained a universal constant with a length dimension"—that is, he believed that one had to give up the concept of strict locality (see Section 1.4.1).

Already in 1928, Bohr had hypothesized that at distances smaller than the so-called classical electron radius, a completely "new physics" takes over, which might be as different from quantum mechanics as quantum mechanics is from classical mechanics. The classical electron radius is 2.8 femtometers (2.8×10^{-15} meters). This is approximately the diameter of an atomic nucleus. Bohr's conjecture turned out to be incorrect: the principles of quantum physics are apparently valid up to the smallest testable distances. Today, most physicists again believe in a new limiting length, but now the decisive distance is no longer the classical electron radius, but the Planck length of 1.6×10^{-20} femtometers. This is a hundred million trillion times smaller than the classical electron radius and correspondingly much more difficult to check (or to disprove). The Planck length is the length at which the quantum effects of gravitation are supposed to become important. We will return to this in more detail in Chapter 7.

In 1943 a theory of a limiting length was not on the horizon, so Heisenberg tried to "isolate those concepts out of the conceptual constructions of quantum theory of wave fields that probably will not be affected by future changes and therefore will also be a part of the future theory." Thus, he did not wish to speculate about what happened at extremely small distances, but instead looked for a theory valid for the case in which the distances involved are large compared to the limiting length. In this "effective" theory, only properties of particles at large distances should be taken into consideration, and the general conservation laws should be satisfied, especially the law of conservation of probability.

Such a theory should be applied primarily to scattering processes, because in that case, particles can be considered to be free long before and long after the scattering process, since they are far apart and cannot interact. The quantum-mechanical probability amplitude for a scattering process is called an S matrix. The approach to elementary particle physics that does without a detailed microscopic dynamics is called S-*matrix theory*.

Christiaan Huygens's derivation of the laws of collision (1669) is a very good historical example of the application of a theory that relies only on general principles. He made no detailed assumptions about the inner dynamics of the bodies colliding with each other but relied only on—expressed in modern language—the conservation laws of momentum and energy. With these assumptions, he was able to derive the laws of elastic collisions; these laws played an important role in the history of mechanics.

Heisenberg soon turned his back on S-matrix theory, and in his last published paper he wrote that dynamics has to be the foundation for our comprehension of particle physics. But because of the difficulties in applying quantum field theory in strong interactions, S-matrix theory gained a significant following in the mid-twentieth century. In his trendsetting book S-*matrix Theory of Strong Interactions*, G. Chew writes, "I do not wish to assert (as does Landau) that conventional field theory is necessarily wrong, but only that it is sterile with respect to strong interactions and that, like an old soldier, it is destined not to die but just to fade away."

3.2 Scattering Amplitudes

Before I proceed to the actual subject of S-matrix theory, I must make some remarks on scattering amplitudes in quantum physics. This section is rather technical, but I do not want to persuade the reader to skip it. (In that case, I would not have written it.) However, I would like to draw the reader's attention to the fact that this material is very rarely needed outside this chapter.

The probability of the occurrence of a process is given by the square of the modulus of a (complex) *probability amplitude*. The phase of this amplitude plays a role only in the superposition of two processes. The fact that a quantum-mechanical amplitude contains information on two numbers, modulus and phase, or alternately real and imaginary parts, can perhaps someday be used in "quantum computers" and may open a new dimension in computation. The special principle of quantum physics that processes are described

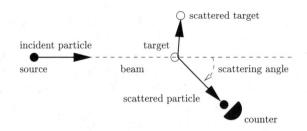

Figure 3.1. Schematic representation of a scattering experiment.

by complex amplitudes but that only the modulus is directly accessible to measurement will, on several occasions, occupy us in what follows.

A *scattering amplitude* is the probability amplitude for a definite scattering process; the square of the modulus is proportional to the cross section. Figure 3.1 displays a scattering experiment schematically.

In the famous scattering experiments of Geiger and Marsden (Section 1.2) and in those of Fermi and his collaborators (Section 2.5) there were only two relevant implied parameters: the energy of the incident particles and the angle of the scattered particles. All other quantities, such as the recoil of the target particle, are fixed by conservation laws. This is no different from what one obtains in classical mechanics. In the experiment of Geiger and Marsden, the energy of the incident alpha particles was determined by the decay energy of the radioactive radon gas—the source of the alpha particles—and the scattering angle was registered for each event. Fermi, however, was able to influence the energy through beam extraction. He also measured only at a fixed scattering angle, which was determined by the position of the counters. In both experiments the target was fixed; in one case it consisted of gold (Geiger and Marsden), and in the other of gaseous hydrogen.

An elastic scattering process is not the only one possible. In scattering of negatively charged pi mesons on protons, one can observe *charge exchange*. The negatively charged meson can change into a neutral one, and the proton into a neutron; both the initial and final states must of course have the same total charge. Another possible reaction is one in which the pi meson and the proton change into a negative K meson and a positive sigma hyperon. Here, in addition to charge conservation, it is also necessary that conservation of strangeness be satisfied (for strong interactions).

If the energy of the incident particles is sufficiently high, additional particles can be produced. For example, a positive pi meson striking a proton can produce two additional mesons, one positively and one negatively charged. In cases in which the number or kind of particles in the final state is different from that in the initial state, one calls the scattering process inelastic. All possible scattering amplitudes form the S matrix, and therefore a scattering amplitude is sometimes called an S-matrix element.

One of the most important properties of the S matrix is a consequence of probability conservation, which means that the total probability that something comes out of a scattering process is 1. This sounds rather trivial, but for the S matrix it has far-reaching consequences. Quite generally, we can consider the S matrix as an operator that transforms the initial state into the final state. The physical requirement of conservation of probability can be formulated mathematically in the following way: the S matrix is a generalized rotation; that is, it conserves lengths and angles in a complex space. Such a matrix is called *unitary*. One of the consequences is the following: if no inelastic process is possible, then the probability of elastic scattering—including that nothing happens at all—is 1. If inelastic processes are possible, then the probability is less than or equal to 1. It is astonishing but true that conservation of probability fixes many properties of the S matrix.

Though energy and scattering angle are the most intuitive parameters for describing a scattering process, they are often not the most convenient ones for a theoretical description. Instead, one may turn to what are called Mandelstam variables (after the physicist S. Mandelstam). They are relativistically invariant, that is, they have the same value in all uniformly moving reference frames. One can show easily that in a scattering process in which two particles scatter without production of additional particles, there are only two independent variables.

In the so-called center of momentum frame (or center of mass frame) two particles—let us call them a and b—hit each other with equal momentum but from opposite directions. Then the total momentum, that is, the sum of the momentum $\vec{p}_a$ of particle a and $\vec{p}_b$ of particle b, is zero. In this frame, one of the Mandelstam variables is the square of the total energy of the incoming particles, called s. The other variable, called t, is connected in a simple manner with the scattering angle.

Generally, one calculates the Mandelstam variables in the following way: one introduces the *four-momentum p* as a combination of energy and the three components of momentum. The variable s is the square of the sum of the four-momenta of the incoming particles, and the variable t is the square of the difference of the four-momenta of an incoming and an outgoing particle. This

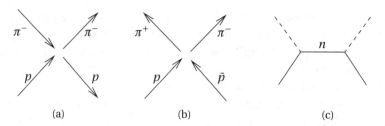

Figure 3.2. (a) The relation between scattering of a negative pi meson (π^-) on a proton (p) (process 1) and (b) the annihilation of a proton and an antiproton ($\bar{p}$) into a pair of pi mesons (process 2). The lines are reversed, and particles are simultaneously changed into antiparticles. (c) The Feynman graph contributes to both processes. The solid outer lines designate a proton or antiproton, the dashed lines a pi meson; the inner solid line is a virtual neutron.

can be expressed more succinctly mathematically. Consider a process in which prior to scattering there are particles a and b, and in the final state particles a' and b'. This is analogous to a chemical reaction expressed in the formula $a + b \rightarrow a' + b'$. The Mandelstam variables are $s = (p_a + p_b)^2$ and $t = (p_a - p_{a'})^2$, where p_a denotes the four-momenta of particle a, p_b that of particle b, and so on.

The consequence of the fact that in a quantum field both particles and antiparticles occur is that incoming particles can be replaced by outgoing antiparticles and vice versa, as indicated in Section 1.4.2. This has far-reaching consequences for the scattering amplitudes: one and the same amplitude can describe several different processes.

As an example we consider two processes: first, elastic scattering of a negative pi meson (π^-) on a proton (p), displayed as process 1 in Figure 3.2(a); and second, process 2, in which a proton and an antiproton ($p\bar{p}$) are annihilated and a positive and a negative pi meson are created, as shown in Figure 3.2(b). We can see from Figure 3.2 that we can transform the representation of process 1 into that of process 2 by changing the outgoing proton into its incoming antiparticle and by changing the incoming negative pi meson into its outgoing antiparticle, that is, a positive pi meson. The same function that describes pi meson proton scattering also describes the annihilation process. For process 1, the variable s is the square of the energy and t is the momentum transfer, while in process 2 the variables have changed their roles: here the variable t is the square of the energy and s is the momentum transfer. This can easily be seen, since the reversal of lines entails a sign change of the four-momenta. The sum

$s = (p_p + p_{\pi^-})$ becomes the difference $(p_p - p_{\pi^+})$, and correspondingly, introducing the notation p'_p to denote the four-momenta of the scattered proton, $(p_p - p'_p)$ becomes $(p_p + p'_{\bar{p}})$.

This is a nice formal result, but it is of use only if we know the scattering amplitude for all values of s and t. One can calculate that for process 1, the variable t has to be negative and s positive, whereas in process 2, it is just the other way around. For those regions of variable values where we can make measurements for process 1, the information cannot be used to make predictions for process 2 and vice versa. So we cannot test the theoretical result directly. But the situation is not completely hopeless. As is often the case, recasting the situation in terms of the complex helps. Here, "complex" is meant in the mathematical sense: one considers not only the amplitudes in the "unphysical region," that is, for those values of s and t that are not accessible to measurements, but one also makes the variables complex; that is, one gives them an imaginary part.

This might at first seem abstruse, but by allowing the variables to take complex values, one can make use of the powerful machinery of the theory of complex functions. This is why complex variables have a long tradition in such practical fields as electrical engineering. Especially important is the concept of analytic continuation. If one has the definition of a well-behaved function in a limited region, it is possible without further information to define it in ("continue" it to) other regions. It is thus possible to predict the annihilation process (process 2) from a precise knowledge of the amplitude for the scattering process (process 1). For that, however, one would have to determine the amplitude for process 1 at infinitely many points with infinite precision. This is impossible, but pragmatically, one can make certain model assumptions with a few free parameters, adjust the parameters by comparison with experiment, and then use this information to make predictions about the other process. This was the basis of a whole new approach to particle physics, which we will discuss in the next section.

The Feynman graph of Figure 3.2 contributes to both reactions. Since similar graphs will be important in the next section, we will discuss the figure in some detail. The internal line of the virtual neutron leads to an infinite value—a *pole*—in the scattering amplitude when the variable s has the value of the square of the neutron mass times the fourth power of the velocity of light: the scattering amplitude is proportional to $1/(s - m_n^2 c^4)$. For process 1, the variable s is the square of the total energy of the initial state; thus it must be at least as large as the rest energy of the two particles: $s > (m_p + m_\pi)^2 c^4$. This is the *threshold energy*; it has the value $(1077.84)^2$ MeV^2. In a real scattering process, the neutron pole at $s = 939.57^2$ MeV^2 can never be reached, since it is below the threshold energy. The same is true for process 2, where in this case the pole is situated at an "unphysical" scattering angle.

An important method for continuing a function from one region of the complex plane into another uses what are called *dispersion relations*. They follow only from the requirement of causality: that an effect not occur before the cause. Such relations were used first in optics in order to derive relations between the refraction and absorption coefficients. In quantum field theory, causality corresponds to locality (see Wightman axiom W4, Section 2.4), and so it is not surprising that in a local quantum field theory, dispersion relations can be proved rigorously. This was first realized by Gell-Mann, Goldberger, and Thirring in 1954. H. Lehmann, K. Symanzik, and W. Zimmermann also gave a proof of it in 1954 and established in particular the relation between quantum field theory and S-matrix theory.

A classic example of the successful application of dispersion relations is the prediction of a new kind of meson, the rho meson, by W. R. Frazer and J. R. Fulco in 1959. They had investigated the dependence of electron–proton scattering on the variable t and came to the conclusion that these data are best explained by the exchange of a meson of spin 1 and mass about 600 MeV/c^2. This meson can be observed directly as a resonance in a system of two pi mesons. We have seen in Section 2.7 that there is indeed a resonance in the system of two pi mesons; Figure 2.13 shows the resonance curve. However, the mass of the observed meson is larger than what was predicted, namely 770 MeV/c^2.

3.3   Bootstrapping and Nuclear Democracy

To "bootstrap" means to pull oneself up by ones own bootstraps. One of the first inventors of this method was the Baron Münchhausen, who pulled himself, together with his horse, out of a swamp—not by his bootstraps, but rather by his pigtail. Lovelace, one of the main developers of this field, therefore jokingly called Münchhausen the inventor of the bootstrap method. The Baron was famous for his fantastic stories, and this was a bad omen, for the bootstrap method turned out to be, if not quite a cock-and-bull story, at least a dud.

What came to be known as the bootstrap program was a search for a radical change in the aim of elementary particle physics. Previously, new structures had been found by digging more deeply into subatomic structures. Matter is composed of molecules and atoms, the atoms consist of a nucleus and an electron shell, and the nucleus comprises neutrons and protons. Then it was discovered that there are many subnuclear particles, appearing only under extreme conditions, such as in cosmic rays or accelerator experiments. Was it sufficient simply to continue to look for further constituents of the subnuclear particles, or should one search for completely new concepts?

The bootstrap program took the second path and assumed that there are no truly "elementary" particles and that all subnuclear particles are on an equal footing, giving rise to the program's other name: nuclear democracy. The name "bootstrap" came from the following hypothesis: instead of assuming the existence of a multitude of particles, it was postulated that these particles mutually entail one another through self-consistency conditions, and thus so to speak pull themselves up by their own bootstraps. A cornerstone of the consistency conditions was the mathematical condition of unitarity of the S matrix and hence the conservation of probability.

If the bootstrap program had been successful, I would be in a difficult situation at this point: I would have to describe theoretical subtleties and delicate technical processes that defy a simple explanation. But since events went in a different direction, I have only my duty as historical chronicler and can therefore be brief. Of course, many of the methods developed in this period remain in the toolbox of every particle physicist, but they do not play the central role that they commanded in the 1960s.

Let us consider—from a purely theoretical point of view—the scattering of two pi mesons. We know that there is a resonance in the two-pi-meson state, namely the rho meson, at an energy of 770 MeV. This is represented graphically in Figure 3.3(a). One can also construct the graph of Figure 3.3(b) for this process of pi meson–pi meson scattering. However, if one simply adds the two diagrams, as is done in perturbation theory, one ends up violating the unitarity of the S matrix.

In the bootstrap program one proceeds as follows: one begins with Figure 3.3(b), calling it the exchange of a rho meson in the t channel. This exchange induces a force between the pi mesons, just as the exchanged photon induces the electromagnetic force, or the exchanged pi meson the nuclear force; see Section 1.4.2 and Figure 1.15.

Using this force, one calculates the scattering of pi mesons, taking into account the unitarity of the S matrix. The result of the calculation should show a resonance in the s channel—Figure 3.3(a)—as a consequence of the force induced by the rho exchange, Figure 3.3(b). There are two free parameters: the mass of the rho meson and its coupling strength to the pi mesons. They are

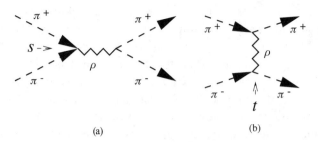

(a)   (b)

Figure 3.3. Contributions of the rho meson to pi meson scattering. The graph in (b) shows that the exchange of a virtual rho meson in the *t* channel leads to an attractive force between the pi mesons that can lead to a resonance in the *s* channel, graph (a), at precisely the energy corresponding to the mass of the rho meson.

fixed by the requirement that the strength and the mass of the calculated resonance in the *s* channel be the same as the input values for strength and mass of the exchanged rho meson in the *t* channel. The system is thus determined by the condition of self-consistency.

One can proceed further by investigating the scattering of pi mesons on rho mesons—evidently this is even more theoretical than the scattering of pi mesons. In this scattering amplitude, the pi meson can occur as a resonance in the *s* channel and as an exchanged particle in the *t* channel. This makes it possible to extend the self-consistency conditions. In this approach, the pi meson and the rho meson are on an equal footing: we can say with equally good—or bad—justification that the rho meson is a bound state of two pi mesons, or that the pi meson is a bound state of a rho meson and a pi meson. Of course, there is a difference between the pi meson and the rho meson in the real world: the pi meson is a particle that lives long enough to make visible traces, whereas the rho meson is a resonance with a calculated lifetime of only three-billionths of a femtosecond. However, this difference is due only to the different masses. Since the mass of the rho meson is twice that of the pi meson, the rho meson can decay into pi mesons but not the other way around.

The first quantitative calculation in the bootstrap program was performed in 1961 by F. Zachariasen. The result was neither a triumph nor a disaster. The mass of the rho meson was computed to be two and one-half pi meson masses, which is less than half of the experimental value. The calculated coupling was about three times the experimental value. So one cannot speak of quantitative agreement, but at least it was shown that the self-consistency condition can lead to numerical results.

Despite great effort, the bootstrap program yielded no truly satisfactory results, and in 1968, D. Atkinson pointed out great theoretical difficulties as well. The program did not die suddenly, but it suffered the fate that Chew had predicted for field theory: it simply faded away. However, the idea of trying new approaches in particle physics had an important influence on the further development of this field, as we shall see in Sections 4.3 and 7.7.

3.4 Rigorous Theorems and Complex Angular Momenta

An unexpected result of the scattering experiments of Fermi and his collaborators, one that was significant for the further development of particle physics, was that the energy dependence of the cross section for scattering of strongly interacting particles (hadrons) was far from uninteresting; it showed a rich resonance structure. It was clear that this had to end at some energy. The higher the energy, the easier and faster the resonances can decay. Because of the energy–time uncertainty principle, this means that they become broader and broader and finally cease to be resonances.

Figure 3.4 shows the total cross section for the scattering of negatively charged pi mesons on protons. One can see clearly that above a total energy of 2.5 GeV, a resonance structure is no longer visible.

This led naturally to the question of how cross sections of hadrons behave at very high energies. It is astonishing that one can give relevant answers to this question without knowledge of the special dynamics of strongly interacting particles. From general principles of quantum field theory and conservation of probability (unitarity of the S matrix) alone, M. Froissart showed (1961) that the total cross section can increase at most with the square of the logarithm of energy; this is the famous *Froissart theorem*. A naive argument leads to the conclusion that at high energies, where resonance phenomena no longer play a role, the cross section is determined by the size of the hadrons. This would imply that at high energies, the cross sections become independent of energy. That this need not be the case is closely related to the possibility of producing particles. The logarithmic increase permitted by theory is weak, but nevertheless, it grows without bound.

More specific than the general Froissart theorem are the following considerations. In the previous section we saw that the exchange of a rho meson contributes to the scattering amplitude. From general considerations, it follows

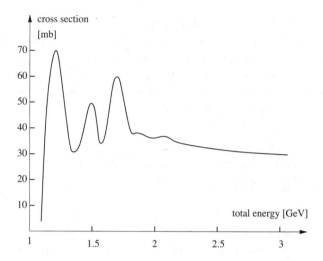

Figure 3.4. Total cross section for scattering of negatively charged pi mesons on protons. The cross section is displayed in millibars (10^{-27} cm^2); the energy is the total energy including the rest masses, and the lowest possible energy is therefore 1.075 GeV. The prominent resonance at 1.232 GeV is the delta resonance.

that the higher the spin of the exchanged (virtual) particle, the faster the cross section increases. A rho meson with its spin of 1 produces a constant contribution to the cross section, but a meson with spin 2 already yields a contribution that increases quadratically with energy. Mesons with higher spin yield an even more rapid growth of the cross section.

But such a fast increase is forbidden by the Froissart theorem, which allows at most a logarithmic increase. This contradiction was resolved in an unexpected way: if one adds the contributions of all mesons, it can happen that the sum shows a much weaker increase with energy than the individual components. This is the result of the so-called Regge theory, whereby not only the energy and momentum transfers are considered to be complex quantities, but the angular momenta as well. Again this may sound rather esoteric, but it is a well-known procedure in mathematical physics, developed originally for calculating such concrete problems as the propagation of radio waves over the ocean.

If the contributions of many mesons with increasing spin are summed, one speaks of the exchange of a *Regge trajectory* in the *t* channel. Whenever this trajectory takes integer or half-integer values—that is, the values allowed for

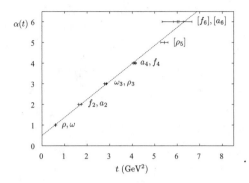

Figure 3.5. A trajectory for mesons, marked by their symbols (ρ, ω, f_2, ...). The variable t is the square of the meson mass in energy units; $\alpha(t)$ is the spin. Mesons occur if $\alpha(t)$ takes integer values.

spin—it represents a meson or a baryon. And indeed, such behavior has been observed. Figure 3.5 shows a trajectory for mesons. Displayed are the masses and spins of mesons with strangeness zero, marked by their symbols as ρ, f_2, etc. The figure shows that to quite a good approximation the trajectory is a straight line. It is important that the line intersects the y-axis at a value not greater than 1, since only then is the Froissart theorem satisfied. As one can see, this is the case for the trajectory in Figure 3.5.

The continuation of angular momentum to complex values was introduced by Regge into quantum-mechanical scattering theory in 1959. In 1962, G. Chew and S. N. Frautschi established the relation to exchanged resonances. The mathematical foundation goes back to G. N. Watson (1918).

The relation between resonances and high-energy scattering is perhaps the most important contribution of the bootstrap program to the development of particle physics. In 1968, G. Veneziano developed a nice model that shows this relation in a mathematically precise way. It was hailed at the time by some bootstrap enthusiasts as the "Maxwell equations of particle physics." That was, to be sure, a rather substantial exaggeration, but the model developed a life of its own. Its importance for present-day particle physics is limited, but it was the starting point for the very speculative development known as string theory, which will be discussed briefly in Section 7.7.

4

Composite Elementary
Particles

The new pathway to particle physics described in the last chapter turned out
to be a dead end, and so old concepts were revived. In the past, the old atom-
istic program had been very successful: many substances could be reduced to
the chemical elements, and the elements reduced to the constituents of atoms
such as electrons, protons, and neutrons. So why should one stop there and
not try to explain the many new elementary particles as composites of a few
fundamental constituents?

4.1   First Attempts

We have seen that the situation in the middle of the twentieth century was am-
biguous. After the discovery of the pi meson, there were good reasons to believe that a giant step had occurred in understanding subnuclear matter, but
it was obvious that the predictions of quantum field theory for strong interactions were far less reliable than those for electrodynamics. Furthermore, the
newly discovered V particles (1947) showed that the situation was less simple
than had been hoped. These V particles, later called strange particles, seemed
to be completely superfluous for understanding ordinary matter.

In 1949, the discovery of new elementary particles encouraged Fermi and
Yang to speculate that the newly discovered particles might be composite. They
investigated whether the pi meson was perhaps a bound state of a nucleon and
an antinucleon. Though at this time neither the antiproton nor the antineu-
tron was an experimentally established particle, they took their existence for
granted. Even more daring was the hypothesis that there exists an extremely

strong force of very short range that binds the nucleon and antinucleon very tightly. The total mass of these two constituents, nearly 1900 MeV/c^2, has to be reduced to the pi meson mass of 140 MeV/c^2. Thus the (negative) binding energy has to be more than 1,700 MeV. Since the typical binding energy of a nucleon inside a nucleus is of order 7 MeV, it was clear that the forces between the nucleons inside a nucleus are insufficient to bind a nucleon and an antinucleon so tightly. Furthermore, the new force should not be induced by the exchange of a new meson, for that would only shift the problem. Fermi and Yang therefore declared, "We will try to work out in some detail a special example more as an illustration of a possible program of the theory of particles, than in the hope that what we suggest may actually correspond to reality."

In their model, one property of pi mesons found a natural explanation: the negative internal parity. Since fermions and antifermions have opposite internal parities, the most natural parity for the bound state is the product of the internal parities of the constituents, and that is negative.

Enrico Fermi (Nobel Prize 1938) was one of the greatest physicists of the twentieth century; I have described his pioneering contributions on several occasions. Chen Ning Yang (Nobel Prize 1957) was a promising young genius. However, in spite of the prominence of Fermi and Yang, the paper in which they explained their idea drew little interest when it appeared. Seven years later, when many more new particles had been discovered, S. Sakata took up the idea and applied it to strange particles. He considered the proton, the neutron, and the lambda hyperon as elementary constituents. The pi meson was for him, as it was for Fermi and Yang, a bound state of a nucleon and an antinucleon, while the K meson was a bound state of a nucleon and an anti-lambda hyperon. This was completely analogous to the Fermi–Yang model.

But there were more strange baryons around: the sigma hyperon (σ-hyperon) with strangeness -1, appearing in three charge states, positive, negative, and neutral; and the xi hyperon (Ξ-hyperon) with strangeness -2, which can be negatively charged or neutral. In the Sakata model, these particles were assumed to comprise three particles: two baryons and one antibaryon. The sigma hyperons comprised a nucleon, an antinucleon, and a lambda hyperon (Λ-hyperon), and the xi hyperons consisted of two lambda hyperons and an antinucleon. The model is represented pictorially in Figure 4.1, where one can see that all observed charge states can be constructed.

It was a nice feature of the model that it reproduced just the observed states, but it was unaesthetic that the hyperons were treated on such a different footing. The masses of the various hyperons agree within plus or minus ten per-

| π-meson | K meson | anti-K meson | Σ-hyperon | Ξ-hyperon |

Figure 4.1. Pictorial representation of the Sakata model, in which the mesons, sigma hyperons, and xi hyperons are composed of nucleons (N), antinucleons ($\bar{N}$), lambda hyperons (Λ), and their antiparticles ($\bar{\Lambda}$).

cent, and therefore it seems natural to combine them into a single group, and not to take one, the lambda hyperon, as elementary and the others as composite. And yet the Sakata model is a good example of the dictum of Francis Bacon that truth derives more easily from untruth than from confusion.

In particular, the Sakata model led to the following important development: M. Ikeda, S. Ogawa, and Y. Ohnuki extended it by bringing symmetry arguments into the picture. They considered the Sakata model in the limiting case in which proton, neutron, and lambda hyperon have the same mass. This is a good illustration of the dependence of physical idealizations on external circumstances. One is ready to treat a mass difference as small if it is small compared to currently attainable energies. In 1959, when Ikeda and his collaborators reconsidered the Sakata model, accelerators existed that could accelerate protons to an energy of 30,000 MeV, compared to which the mass difference between the neutron and the lambda hyperon (176 MeV/c^2) is small. In the 1940s, when the attainable energy was only a few hundred MeV, the idealization would have been accepted less readily.

Since the masses of the proton, neutron, and lambda hyperon are equal in the limit under consideration, the masses of the composed mesons have to be equal too. The authors made the very important further observation that one could construct not only the seven mesons then known, namely the three pi mesons, the two K mesons, and the two anti-K mesons, but also an eighth, light, meson. It should be chargeless, like the neutral pi meson, but have isospin 0. They called the predicted particle pi-0' ($\pi^{0'}$). The unsatisfactory situation in the baryonic sector was not altered by these symmetry considerations.

The predicted meson was found about two years later. With a mass of 547 MeV/c^2 it is only 10% heavier than the K meson. The discoverers did not, however, mention the theoretical prediction and called it the eta meson (η-meson). Perhaps the unconvincing classification of the baryons was to blame for the Sakata model not having been taken seriously.

Fermi and Yang could well have ventured such a prediction. From a nucleon and an antinucleon one can construct four different mesons: the three charge states of the pi meson with isospin 1 and a fourth one with isospin zero. The construction is the same as that presented in Figure 1.23, except that one of the nucleons has to be replaced by an antinucleon. In 1949, however, only the three charge states of the pi meson had been found, and the K mesons had not even been identified. Fermi and Yang certainly knew enough group theory to draw such a conclusion, but at that time, additional mesons were not yet on the horizon.

Perhaps it is good that Fermi and Yang did not make such a prediction, since they would necessarily have concluded that the mass of the predicted meson should be nearly equal to the mass of the pi meson. Only later, with more powerful accelerators, was one prepared to combine particles of 140 MeV and over 500 MeV/c^2 into a single class of light mesons.

In a subsequent paper, Ikeda, Ogawa, and Ohnuki investigated the mathematical structure of the symmetry corresponding to the physical equivalence of the proton, neutron, and lambda hyperon; they found it to be a unitary group of dimension 3, called SU(3). This brings us to another excursion into group theory.

4.2   The Eightfold Way

The orientation of the isospin, that is, its projection on the three axes in an abstract isospace, I_3, and the strangeness S determine the charge of a particle. In Section 2.2 we became acquainted with the formula of Gell-Mann and Nishijima in which the charge Q, isospin orientation I_3, baryon number B, and strangeness S are related by

$$Q = I_3 + \tfrac{1}{2}B + \tfrac{1}{2}S.$$

The isospin symmetry is a consequence of invariance under transformations in an abstract space. It seemed promising to look for symmetry transformations in an even more abstract space containing strangeness and isospin as dimensions. The analysis of the Sakata model by Ikeda and his collaborators had shown that the group SU(3) is the appropriate symmetry transformation group.

The structure of the group SU(3) is very similar to that of SU(2), but the matrices of the simplest representation are here of dimension 3×3, not 2×2. There

triplet {3}	I	I_3	S	antitriplet $\{\bar{3}\}$	I	I_3	S
u	$\frac{1}{2}$	$+\frac{1}{2}$	0	$\bar{u}$	$\frac{1}{2}$	$-\frac{1}{2}$	0
d		$-\frac{1}{2}$	0	$\bar{d}$		$+\frac{1}{2}$	0
s	0	0	-1	$\bar{s}$	0	0	$+1$

Table 4.1. The fundamental triplet (u, d, s) and antitriplet $(\bar{u}, \bar{d}, \bar{s})$ of SU(3). Here I is the isospin, I_3 the isospin orientation, and S the strangeness.

are two independent triplets transforming under these matrices; one is called the fundamental triplet {3}, the other the fundamental antitriplet $\{\bar{3}\}$. In the Sakata model, the fundamental triplet corresponds to the proton, the neutron, and the lambda hyperon; the antitriplet, to the respective antiparticles. The isospin group SU(2) is contained in the group SU(3). Mathematically speaking, SU(2) is a subgroup of SU(3). Therefore, the elements of the fundamental triplet can be analyzed according to their isospin. The proton and the neutron form an isospin doublet, and the lambda hyperon is a singlet, which is invariant under isospin transformations.

I will not go into details, but only describe a simple construction scheme for all representations of SU(3). It is analogous to the construction for SU(2) displayed in Figure 1.23. We will free ourselves from the Sakata model, and for reasons that will become clear later, we will designate the three components of the fundamental triplet by the letters u, d, and s, which stand for *u*p, *d*own, and *s*trange.

The isospin and strangeness of the elements of the triplet and antitriplet are indicated in Table 4.1. If in a plane we display the strangeness from bottom to top, and the isospin orientation I_3 from right to left, we obtain for the fun-

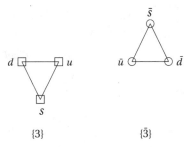

$\{3\}$ $\{\bar{3}\}$

Figure 4.2. The fundamental triplet {3} of the group SU(3) and the corresponding antitriplet $\{\bar{3}\}$. The strangeness S is displayed from bottom to top, the isospin orientation I_3 from left to right.

damental triplet a triangle standing on its vertex, while for the antitriplet, the triangle stands on its base; see Figure 4.2.

Now let us form bound states of the particles that correspond to the members of the fundamental triplet and antitriplet. In the language of group theory, we form the product $\{3\} \times \{\bar{3}\}$ of the representations. Graphically, this is done in the following way: we superimpose the two corresponding triangles in such a way that each vertex of the triangle $\{3\}$ is touched at least once by a vertex of the other triangle, $\{\bar{3}\}$. We thereby arrive at the hexagon of Figure 4.3. If we identify u with the proton, n with the neutron, and s with the lambda hyperon, we obtain the construction of the mesons in the Sakata model. All vertex points are touched by two triangles; that is, each represents a meson as a bound state of two particles. The central point, however, is touched by six triangles; that is, it corresponds to three mesons. One of them is the neutral pi meson, and one is the neutral meson with isospin 0 predicted by Ikeda and his collaborators. Along with the mesons at the corners, these form an eight-dimensional representation of SU(3): an SU(3) octet. The remaining neutral particle is an SU(3) singlet. It transforms according to the trivial representation, remaining unchanged under SU(3) transformations.

That a product of the fundamental representations $\{3\}$ and $\{\bar{3\}}$ yields an octet and a singlet is expressed as

$$\{3\} \times \{\bar{3}\} = \{8\} \oplus \{1\},$$

a deceptively simple-looking equation.

After the discovery of the eta meson (η, formerly $\pi^{0'}$), the classification of the mesons as an octet was very satisfactory, especially since it became more

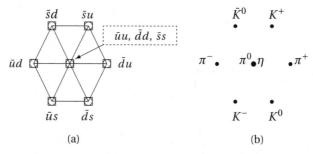

(a)   (b)

Figure 4.3. (a) Octet representation of the elements of the group product of the fundamental triplet and antitriplet, $\{3\} \times \{\bar{3}\}$. Each point is touched at least once by each of the triangles $\{3\}$ and $\{\bar{3}\}$ from Figure 4.2. (b) The corresponding pseudoscalar mesons.

and more evident that all members have the same internal parity, -1, and spin 0. Moreover, additional mesons with spin 1 were discovered, which were excellent candidates for a further octet, this time consisting of mesons with spin 1 (vector mesons). In particular, the mesons for the doubly occupied central place had been discovered: the rho mesons and the omega mesons (ω-mesons). This classification, which even allowed predictions, could be counted as a big success for the Sakata model.

But the model failed totally in the case of baryons. The baryons were composed of three copies of the fundamental representation, in group-theoretic language a product of the representations $\{3\} \times \{3\} \times \{\bar{3}\}$. From the known properties of the baryons, it was evident that none of them could be accommodated in the resulting representations. Independently, M. Gell-Mann and Y. Ne'eman, as well as Speiser and J. Tarski, proposed to abandon the Sakata model. They combined the baryons, like the mesons, into an octet representation. The elements of this octet are the nucleons and the sigma, lambda, and xi hyperons, collected in Table 4.2 and displayed graphically in Figure 4.4(a); the hypercharge Y, the sum of baryon number and strangeness, is displayed from bottom to top, the isospin orientation I_3 from left to right. The neutral sigma hyperons and the lambda hyperons have hypercharge $Y = 0$ and isospin orientation $I_3 = 0$, and thus occupy the central place.

Gell-Mann called this symmetry the *eightfold way* (an allusion to the Eightfold Path of Buddhism). Today it is called *flavor*-SU(3). The assignment seems compelling, but that was not the case at the time it was proposed. The group SU(3) and three other possible symmetry groups are treated with equal weight in a review article from 1962. There were several reasons for this. The prop-

baryon	symbol	I	I_3	S	Y	mass (GeV/c^2)
proton	p	$\frac{1}{2}$	$+\frac{1}{2}$	0	$+1$	938.27
neutron	n		$-\frac{1}{2}$	0	$+1$	939.57
Λ-hyperon	Λ	0	0	-1	0	1115.68
positive Σ-hyperon	Σ^+		$+1$	-1	0	1189.37
neutral Σ-hyperon	Σ^0	1	0	-1	0	1192.64
negative Σ-hyperon	Σ^-		-1	-1	0	1197.45
neutral Ξ-hyperon	Ξ^0	$\frac{1}{2}$	$+\frac{1}{2}$	-2	-1	1314.83
negative Ξ-hyperon	Ξ^-		$-\frac{1}{2}$	-2	-1	1321.31

Table 4.2. The octet of baryons; I is the isospin, I_3 its orientation, S the strangeness, and $Y = B + S$ the hypercharge.

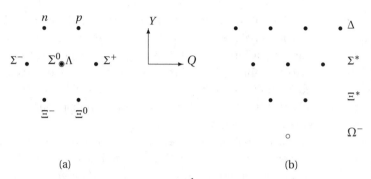

(a)   (b)

Figure 4.4. (a) The octet of baryons with spin $\frac{1}{2}$ and (b) the decuplet of baryons with spin $\frac{3}{2}$. The hypercharge is displayed from bottom to top, the isospin orientation I_3 from left to right. The omega hyperon (Ω^-, open circle) was a prediction of the symmetry.

erties of the baryons were not so well known, and furthermore, the symmetry breaking is rather strong. The mass difference between the nucleon and the delta resonance, which is not a member of the octet, is smaller than that between a nucleon and a xi hyperon, which are both octet members.

M. Gell-Mann and S. Okubo made plausible assumptions about symmetry breaking and derived relations among the masses of elements of the same representation, whereby m_N, m_Ξ, m_Λ, and m_Σ—the masses of the nucleon and xi, lambda, and sigma hyperons—are related by

$$2(m_N + m_\Xi)/(m_\Lambda + 3m_\Sigma) = 1;$$

the experimental result is 0.96.

The eightfold way gained full recognition through a prediction that was regarded as rather spectacular at the time. In 1962, resonances with strangeness 0, 1, and 2 were known. Their masses ranged from 1,232 to 1,530 MeV/c^2. It was impossible to accommodate them in an octet representation, since such a representation has no place for a delta resonance with isospin $\frac{3}{2}$. There was, however, the possibility of combining them in a ten-dimensional representation, a decuplet. In this case there was one resonance too many and one missing. We will return to the superfluous resonance later. The missing one should have strangeness -3 and isospin 0; the mass could be predicted to be about 1,680 MeV/c^2. This particle, called the omega hyperon (Ω-hyperon), was found in 1964. The experimental mass was very close to the predicted one, 1,672 MeV/c^2. The first picture of the omega hyperon, in a two-meter hydrogen bubble chamber, is shown in Figure 4.5.

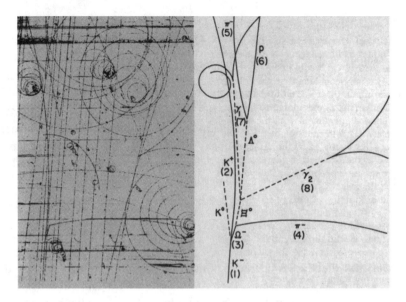

Figure 4.5. Detection of the omega hyperon. In a bubble chamber filled with liquid hydrogen, a positively charged K meson hits a proton and through the strong interaction produces an omega hyperon (Ω^-) and two K mesons—one positively charged (K^+) and one neutral (K^0). The omega hyperon decays via the weak interaction into a xi hyperon (Ξ^0) and a negatively charged pi meson (π^-). The xi hyperon decays further into a lambda hyperon (Λ^0) and two photons (γ_1, γ_2). Each photon creates an electron–positron pair. The lambda hyperon decays into a proton and a negative pi meson (p and π^-). The invisible traces of neutral particles are indicated in the schematic drawing by dashed lines; the neutral particles are identified by their decay products.

In addition to the prediction of SU(3) symmetry, the discovery of the omega hyperon also confirmed another conjecture, namely, that there is no fundamental difference between stable particles and resonances. The omega is stable under strong interactions. It lives long enough to leave a visible trace in a bubble chamber, as can be seen in Figure 4.5. The reason for its stability is that the lightest state into which the omega hyperon could decay through strong interactions consists of a negative xi hyperon and a neutral K meson. This state has the same baryon number, the same strangeness, and the same charge as the omega, but at $1{,}819\ \mathrm{MeV}/c^2$ its total mass is much larger than that of the omega ($1{,}672\ \mathrm{MeV}/c^2$). The actual decay modes (mainly into a lambda hyperon and a negative K meson) violate strangeness conservation and therefore proceed only via weak interactions. The fact that the same representation contained stable particles and unstable resonances was of course confirmation for the "nuclear democrats," who considered all strongly interacting particles to be on the same footing. I do not want to elaborate here on the other successes of this symmetry, other than to note that it had very important applications for weak interactions. We will return to this topic in Section 6.3.

One can trace a direct line from the original paper of Fermi and Yang to the eightfold way by way of the Sakata model. Along that line, the original idea to consider some particles as composite and some as elementary moved into the background. One thing, however, became clear: it was not economical to consider the baryons as elementary particles.

4.3   The Quark Model

"If we assume that the strong interactions of baryons and mesons are correctly described in terms of the broken 'eightfold way,' we are tempted to look for some fundamental explanation for the situation." This is the first sentence of the paper entitled "A Schematic Model of Baryons and Mesons," published by Gell-Mann in the journal *Physics Letters*. It was submitted on January, 4, 1964, and published in February, in the first issue of the same year. On January 13, 1964, there appeared a CERN preprint by George Zweig, "An SU(3) Model for Strong Interaction Symmetry and its Breaking." The two papers ushered in a new era of particle physics. Although their motivations and scopes were quite different, the results agreed in essential points.

Before we come to physics proper, let us take a brief historical digression to explain why Zweig's important paper was never published in a scientific journal. The policy of CERN was to publish only in European journals. George Zweig, who had just received his degree from the California Institute of Technology (where Gell-Mann was a professor), wanted to publish this paper in the American journal *Physical Review*. He argued that he had received no financial support from CERN and that his scholarship from the United States even included a grant for CERN. After a number of lively discussions with the head of the CERN theory division, Zweig's paper did not see publication in a journal.

Though Zweig's preprint appeared a few days later than Gell-Mann's paper, I will begin with the former. Zweig's arguments are more intuitive than those of Gell-Mann. Zweig took the Sakata model very seriously. He was strengthened in his conviction by the discovery of a new meson, the phi meson (ϕ-meson), at the Brookhaven National Laboratory in 1963. This particle had spin 1, isospin

meson	S	I	mass MeV/c^2	width MeV/c^2	dominant decay particles	dominant decay fraction	E_{rest} MeV
ρ	0	1	771	149	$2\,\pi$	ca. 100 %	500
ω	0	0	782	8.44	$3\,\pi$	ca. 90 %	370
ϕ	0	0	1019	4.26	$2\,K$	83 %	29

Table 4.3. The three vector mesons that inspired Zweig. Here S is strangeness, I isospin, and E_{rest} the kinetic energy of the decay products.

0, and strangeness 0. Thus, it had the same quantum numbers as the already known omega meson; but it was not only heavier, 1,020 versus 782 MeV/c^2, its lifetime was twice that of the omega meson; see Table 4.3.

This was unexpected, since the heavier the particle, the more easily it can decay. The decay channels are also different: whereas the omega meson decays mainly into three pi mesons, the phi meson decays into a K meson and its antiparticle. This is mysterious. The sum of the rest energies of the two K mesons is nearly the rest energy of the phi meson, and therefore only very little is left for the kinetic energy of the decay products. Thus for the phi meson as well, the three-pi-meson decay should be dominant.

While the eightfold way anticipates two mesons with the same quantum numbers, the Sakata model can explain three kinds of vector mesons (mesons with spin 1) with charge and strangeness zero, as shown in Figure 4.6. Two of the vector mesons consist of nucleons and antinucleons, and one consists of the lambda hyperon and its antiparticle. With that, we can give the following interpretation for the three vector mesons: the two light ones, the omega and the neutral rho meson, are a superposition of nucleon–antinucleon states, while the phi meson is composed of the lambda hyperon and its antiparticle. This explains why the rho and omega meson have approximately the same mass, 770 and 782 MeV/c^2, and why the phi meson, composed of the heavy lambda hyperons, is heavier.

The odd decay mode could also be explained. Apparently, constituents of the mesons can annihilate during decay only with difficulty, and therefore

Figure 4.6. The construction of the neutral rho, omega, and phi mesons in the Sakata model.

the lambda hyperon and its antiparticle must appear in the decay products of the phi meson. This explains the energetically disfavored decay mode into K mesons, since in the Sakata model, the K mesons contain the lambda and antilambda hyperons, respectively; see Figure 4.1.

The rule that constituents of a hadron annihilate each other only with difficulty is a very important empirical observation, called the OZI rule, after Okubo, Zweig, and Iizuka; even today its dynamic foundation is not fully clear.

There were thus some indications that the Sakata model for mesons had predictive power, but there was a problem with the baryons, where it did not work at all. Here Zweig made a revolutionary proposal. He suggested that the baryons are composed of three equivalent constituents that form a three-dimensional representation of SU(3), that is, a fundamental triplet. Zweig called these constituents "aces." Gell-Mann, who made the same proposal at practically the same time, was inspired more by literature than the card table. He called them "quarks," after a line from James Joyce's *Finnegans Wake*: "Three quarks for Muster Mark." This name has prevailed, and in the following I shall always use the term quarks, even if I deal with Zweig's aces.

The group-theoretic foundation for the formation of baryons out of three fundamental triplets is the following: if one couples three fundamental triplets, one obtains a singlet, two octets, and a decuplet in the formula

$$\{3\} \otimes \{3\} \otimes \{3\} = \{1\} \oplus \{8\} \oplus \{8\} \oplus \{10\}.$$

It is remarkable that just those three representations appear that are experimentally observed: the eight-dimensional one for the "usual baryons" and the ten-dimensional one for the "resonances" (see Figure 4.4). Even for the one-dimensional (trivial) representation there was a candidate, the previously mentioned resonance at $1,405$ MeV/c^2 with strangeness -1, which had found no place in the decuplet.

The isospin and strangeness of the elements of a fundamental triplet are fixed by the isospin and the strangeness of the known baryons. This is depicted in Figure 4.2, where the currently used names are introduced: u quark (u for up, isospin $+\frac{1}{2}$), d quark (d for down, isospin $-\frac{1}{2}$), and s quark (s for strange, isospin 0, strangeness -1). If the three constituents of the baryons are really to be treated on the same footing, then they must have equal baryon number, namely $\frac{1}{3}$. A result that was disliked by many physicists—the charge of the quarks had to be fractional-follows from the fractional baryon number using the Gell-Mann–Nishijima formula, $Q = I_3 + \frac{1}{2}B + \frac{1}{2}S$. The u quark has charge $+\frac{2}{3}$, while the d and s quarks have charge $-\frac{1}{3}$.

baryon	mass (GeV/c^2)	quark content	meson	mass (GeV/c^2)	quark content
p	938.27	uud	K^+	493.68	$u\bar{s}$
n	939.57	udd	$\bar{K}^0$	497.65	$d\bar{s}$
Λ	1115.68	uds	η	547.75	$s\bar{s}\,(q\bar{q})$
Σ^+	1189.37	uus	π^+	139.57	$u\bar{d}$
Σ^0	1192.64	uds	π^0	134.98	$q\bar{q}\,(s\bar{s})$
Σ^-	1197.45	dds	π^-	139.57	$d\bar{u}$
Ξ^0	1314.83	uss	K^0	497.65	$s\bar{d}$
Ξ^-	1321.31	dss	K^-	493.68	$s\bar{u}$

Table 4.4. Masses and quark contents of the baryon and pseudoscalar SU(3) octet; q stands for u or d, while a bar designates an antiquark; for example, $\bar{u}$ is the anti-u quark.

We already have seen in Section 1.5.2 that spins have to be integer or half-integer. The baryons, consisting of three quarks, have spin $\frac{1}{2}$, and therefore the spin of the quarks must also be $\frac{1}{2}$. No half-integer spin could be constructed from three integer spins; the quarks are thus fermions. Mesons that are constructed of a quark and an antiquark must have integer spin, since two spins of $\frac{1}{2}$ always combine into an integer spin. The quark model therefore explains why all mesons have integer spin and all baryons have half-integer spin.

All results from the Sakata model for mesons are still valid. One has only to replace the proton by the u quark, the neutron by the d quark, and the lambda hyperon by the s quark.

In an impressive effort, G. Zweig drew a number of conclusions from the model. In particular, he assumed that the breaking of SU(3) symmetry is due only to the different masses of the quarks. If one assumes that the mass of the d quark is some MeV greater than that of the u quark, one can qualitatively understand the masses of the hadrons inside an isospin multiplet, as can be seen from Table 4.4. The negative sigma hyperon, for example, consists of two d quarks and one s quark, and the neutral one of one d, one u, and one s quark, while the positive one contains no d quark at all, only two u quarks and one s quark. Correspondingly, the negatively charged sigma is the heaviest and the positively charged one is the lightest.

The same holds for strangeness, as can be seen from the table. If the mass of the s quark is heavier by 100 to 150 MeV/c^2 than that of the u or d quark, this qualitatively explains the mass differences of the hadrons inside an SU(3) multiplet. The quarks' masses and their inherent problems will be discussed in more detail in Section 6.9.

Though much of the data that is today well established was unknown in 1964 and some factors that were not yet well measured did not meet expectations, the conclusion drawn by Zweig in his great paper is still valid: "The scheme we have outlined has given, in addition to what we already know from the Eightfold way, a rather loose but unified structure to the mesons and baryons. In view of the extremely crude manner in which we have approached the problem, the results we have obtained seem somewhat miraculous."

I come now to the paper of Gell-Mann and his motivation to introduce quarks. In 1962, he published a paper in which he investigated mainly weak and electromagnetic properties of hadrons. Note that hadrons, in contrast to leptons and photons, interact strongly, but they are also subject to weak and electromagnetic interactions. In his paper, Gell-Mann derived some important properties "from a formal field theory model based on fundamental entities from which the baryons and mesons are built up." This formal field theory model was known as the *current algebra* method, which was developed further by S. Fubini and his collaborators. It played an important role in particle physics during the late 1960s and early 1970s. It held the banner of field theory high, even in the time of nuclear democracy. Now it is all wrapped up in standard model of particle physics and in so-called chiral models. Therefore, I will only briefly sketch some essential points of this method.

For example, let us consider the electromagnetic current. In quantum electrodynamics it is composed of field operators for the electron. It seemed obvious to transfer this approach to the proton. It is said that in 1932, Pauli made fun of Stern because Stern wanted to measure the magnetic moment of the proton in what was for the time a rather extravagant experiment. Pauli thought that the electromagnetic current of fermions was well understood and therefore the magnetic moment as well. To Pauli, Stern's experiment did not seem to be a high priority, but Stern's curiosity was justified, for he found that the magnetic moment was very different than what was expected from the Dirac theory. It was the first indication that the proton is not as elementary as the electron.

Gell-Mann had introduced abstract current operators for the electromagnetic current and analogous weak currents of hadrons, and had postulated symmetry relations directly for these current operators. In order to derive the symmetry relations for mesons, he had assumed that the current operators of the hadrons are not directly composed of hadron creation and annihilation operators, but that they consist of operators of the above-mentioned fundamental quantities, which transform like a fundamental triplet under SU(3). The selection was oriented according to the Sakata model.

In applying the method of current algebra to baryons, Gell-Mann encountered the well-known troubles of the model. Therefore he assumed—like Zweig in his less-sophisticated approach—that the currents for baryons are composed of operators for new objects that have not been observed and that have baryon number $\frac{1}{3}$. He was more concerned than Zweig about the fractional charges. Since such had never been observed, he suggested that the field quanta of these operators, the quarks, cannot occur as single particles but only in combinations for which the resulting charge is an integer. Thus the credo of the time that all hadrons are to be treated on an equal footing was not violated, at least for the observable particles.

Gell-Mann did not want to enter into a discussion about the reality of unobservable quantities. In his original paper he called the quarks "mathematical objects." He wrote, however, "It is fun to speculate about the way quarks would behave if they were physical particles of finite mass (instead of purely mathematical entities as they would be in the limit of infinite mass)." He was convinced that quarks cannot be observed as free particles. This becomes clear from the last sentence of his paper: "A search for stable quarks ... at the highest energy accelerators would help to reassure us of the non-existence of real quarks." Since one had no idea about the interaction of quarks, but could draw useful information for the current algebra from the quark model, the gourmet Gell-Mann made the much-quoted comparison with a French recipe: cook a pheasant between two pieces of veal; when the pheasant is done throw away the veal (presumably to be eaten by the kitchen staff). The pheasant was the theory of current algebra; the veal, the quark model.

Because of Gell-Mann's quote comparing the quark model to an ingredient to be discarded, which seemed to confirm his disregard for the model, he was reproached for not having taken quarks seriously. In his personal recollections he says, "I did not want to call such quarks "real" because I wanted to avoid painful arguments with philosophers about the reality of permanently confined objects. In view of the widespread misunderstanding of my carefully explained notation, I should probably have ignored the philosophical problem and used different words." We will return to the question of the reality of quarks in Chapter 8. I think that Gell-Mann was totally correct about the cautious and carefully explained notation in his paper.

On the other hand, Zweig's realistic approach had a great impact. He had found that the "naive" quark model could qualitatively explain a large number of remarkable phenomena. That convinced many physicists, especially experimentalists, of the quark model and gave the field an enormous inner dynamic. In the end, however, Gell-Mann turned out to be right, and all the experiments attempting to find free quarks in high-energy accelerators or in cosmic rays only "helped to reassure us of the non-existence of real quarks."

The noninteger quark charges were disturbing to Gell-Mann, but for experimentalists they were a true bonanza. The ionization density of highly energetic particles is practically independent of mass and energy, depending only on the charge. Therefore, it is easy to identify tracks of particles with charges of one-third or two-thirds of the elementary charge. One has only to look for tracks that are much thinner (fewer ions per unit length) than traces of normal particles. There were rumors that fractional charges had been observed, and some old experiments contradicting Millikan's oil droplet experiment were dusted off. But no such rumor has survived to the present, and it is generally accepted that quarks do not exist as free particles.

The quark model was extended by including spin in the symmetry. In doing so, one was taking up an old idea of Wigner who had unified the (usual) spin with the isospin in the group SU(4). The group SU(3) together with the spin group SU(2) yields the unified symmetry group SU(6). If the quarks are the elements of the simplest representation of SU(6), one obtains for the mesons the observed multiplets, an octet of pseudoscalar mesons, an octet of vector mesons, and a vector meson in a singlet, grouped together into a 35-dimensional representation of SU(6).

From three quarks one can construct, among others, a 56-dimensional representation consisting of an octet of spin-$\frac{1}{2}$ baryons and a decuplet of spin-$\frac{3}{2}$ baryons. These are exactly the lightest observed baryons; see Figure 4.4. An astonishing prediction was that the ratio of the magnetic moments of the proton and the neutron should be $-2/3$, very close to the experimental value -0.685. In ambitious projects, attempts were made to construct a dynamic theory out of the static SU(6), which led to the introduction of 35 momentum components. That such an enterprise was doomed to failure was shown later in a theorem of S. Coleman and J. Mandula, to which we will return in Section 7.4.

One of the basic ideas of nuclear democracy, namely, that all baryons "have equal rights," was corroborated by the quark model. Another important concept, duality, could be nicely visualized in the quark model. Duality means that the dynamically exchanged particles are equivalent to the bound states and the resonances. Thus, the two contributions to the scattering of pi mesons, which are displayed in Figure 3.3, must not be added, but they imply each other.

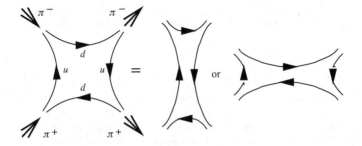

Figure 4.7. The graphs of Figure 3.3 in the quark model.

If one represents the meson as a bound state of a quark and an antiquark and draws the quarks as solid lines, then the two graphs in Figure 3.3 can be represented by the single graph of Figure 4.7. For understanding this dual diagram, it is essential to remember that the ingoing line can represent both incoming particles and outgoing antiparticles. Therefore, the outgoing pi meson at the upper-right corner, for example, consists of an outgoing d quark and an outgoing anti-u quark ($\bar{u}$). The inner lines can be interpreted either as an exchanged rho meson in the t channel or as a rho meson resonance in the s channel.

4.4   The Quarks Assume Color

Physicists who were investigating possible theoretical foundations of the quark model were concerned primarily about two things: the fractional charges and a possible violation of the Pauli exclusion principle. We have seen that Gell-Mann was so concerned about the fractional charges that he accepted quarks as constituents of the hadrons but not as isolated physical particles. The problem with the Pauli principle was more delicate, since it concerned quantum fields in general and was not bound to physical particles. I have mentioned this principle several times. It is the reason for the shell structure of atoms and atomic nuclei. Therefore, the chemical properties of elements are a consequence of this principle. It turned out that it also plays an important role for the structure of hadrons.

The exclusion principle follows from one of the most astonishing results of axiomatic field theory: the theorem of spin and statistics. This theorem was mentioned in Section 2.4. According to this theorem, states of several spin-$\frac{1}{2}$

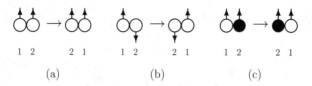

Figure 4.8. The exclusion principle as a consequence of the theorem of spin and statistics. (a) If two particles (fields) in a state agree in all their properties, exchange of the two leads to an identical state. The theorem of spin and statistics proposes that under interchange of two fermions, the state has to change sign. This is impossible for identical states, and therefore a state in which all fermions (fermionic fields) have the same quantum numbers is forbidden by the theorem. (b), (c) Here the two particles (fields) have different properties, and therefore the state resulting from exchange of the two components is not identical to the original one. These two states, (b) and (c), are allowed by the theorem.

particles (fermions) are antisymmetric. This means that the state changes its sign if two components are exchanged. From this it again follows that in a state of two or more particles, no two of them can have exactly the same properties; they have to differ in at least one respect, such as spin, isospin, or momentum.

This can be seen in the following way (see Figure 4.8): let us assume a state consists of two particles and that both agree in all their properties. If we exchange the particles, the state remains unchanged, since the properties are identical, and I cannot even notice that such an exchange has taken place. On the other hand, the theorem of spin and statistics requires that the sign of the state change. This is a contradiction, and therefore such a state cannot exist. This is the content of the exclusion principle. It forbids, for instance, all the electrons of an atom being in the ground state, since in that case they would all have the same properties.

Since quarks are fermions, they should also obey the exclusion principle. For mesons, the exclusion principle yields no restrictions, since they are composed of different particles, namely, quarks and antiquarks (corresponding to Figure 4.8(c)). But there are implications for the baryons. Let us consider the doubly charged delta resonance Δ^{++}. It is in the SU(3) decuplet (Figure 4.4) and consists of three u quarks. If it has the maximal possible spin orientation $I_3 = +\frac{3}{2}$, then the most natural assumption is that all three quarks have the same spin orientation $+\frac{1}{2}$ and also that the spatial distributions of all three quarks have the same properties.

However, this assumption is in contradiction to the exclusion principle. All properties of the three quarks are equal: charge, isospin, orientation, and spatial distribution. A "natural" assumption is of course not absolutely compelling, and one can imagine interactions that bring the delta resonance into accord with the exclusion principle. It turned out, however, that the rather successful symmetry SU(6), mentioned in the previous section, also requires states that are in contradiction to the exclusion principle, and so many physicists were engaged with the problem. They dug out a concept developed in the 1940s, called *parastatistics*. This approach amounts essentially to the introduction of an additional property by which the antisymmetry of a state can be achieved and thus be made compatible with the exclusion principle. I do not want to deal with the many suggestions offered in the 1960s, but only with the one proposed in 1965 by M. Y. Han and Y. Nambu; it was the one that finally led to a field theory of strong interactions.

Han and Nambu assumed that each quark has, besides the properties already known, an additional property, which later was called *color*. It has nothing in common with the properties of light other than the name. According to Han and Nambu, each quark can occur in three different color states, which transform as a fundamental representation of SU(3).

The resulting symmetry, called color-SU(3), is completely new and has nothing to do with the broken SU(3) symmetry of the eightfold way. By introducing this new symmetry, Han and Nambu were able to kill two birds with one stone: they could avoid the problems with the exclusion principle while also avoiding the introduction of fractional charges for the quarks. I will not dwell on the fractional charges, since today quarks with fractional charge are the generally accepted concept. But the additional property of color turned out to be decisive for the further development of quantum field theory in particle physics.

I will use the names that were coined later and always refer to the (broken) SU(3) symmetry of the eightfold way as flavor-SU(3) and to the new symmetry introduced by Han and Nambu as color-SU(3). With these designations we can summarize the situation: the quarks have, besides the three possible "flavors" u, d, and s; additional "colors" called, for example, blue, white, and red.

For example, the u quark appears in three color states: u_{blue}, u_{white}, and u_{red}. With the additional property of color, the doubly charged delta resonance, consisting of three u quarks with all spins up, is no longer forbidden by the exclusion principle. The three quarks can now differ in color; see Figure 4.8(b).

Han and Nambu assumed, like Gell-Mann and Zweig, that the baryons consist of three quarks and the mesons of a quark and an antiquark. They additionally assumed that there is a very strong force that ensures that the lowest-lying

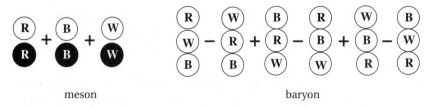

meson   baryon

Figure 4.9. Color composition of a meson and a baryon in the Han–Nambu model. The open circles designate red, blue, or white quarks; the closed circles are the corresponding antiquarks.

states, that is, the observed hadrons, are invariant under transformations of color-SU(3). This means that the observed hadrons are color singlets. One also says that the hadrons are color-neutral. For the baryons, the color-neutral state is automatically antisymmetric, and therefore it must be symmetric under interchange of the already known properties. This is just the above-mentioned "natural" assumption, which now is in accord with the exclusion principle. The color composition of a meson and a baryon is displayed in Figure 4.9.

Han and Nambu also gave hints for a dynamic model that gives rise to the property that the low-lying states are color singlets and that the color octets have high energy. They proposed the exchange of a vector field that is a singlet under flavor-SU(3) but an octet under color-SU(3), which means that this vector field can carry eight different colors.

5

On the Path to the Standard Model

We can summarize the situation in particle physics at the end of the 1960s as follows. Experimental methods had made immense progress since the middle of the century, but theory had failed to keep up—only purely electromagnetic phenomena could be described in a truly satisfactory way. There was a very successful descriptive field theory for the weak interactions, but it was not satisfactory, since it was not renormalizable. The situation was even worse in the realm of the strong interactions. The dream that a completely new approach would be provided by the bootstrap method failed to bear fruit, and the quark model, which did describe some phenomena quite well, was far from being a consistent theory. Great progress had been made in the classification of particles through symmetry considerations, but a dynamic foundation for the symmetries was lacking.

But then, progress was made in the 1970s, and today we have a consistent quantum-field-theoretic description of the weak, electromagnetic, and strong interactions, the *standard model* of particle physics. This success came about as physicists increasingly directed their attention to a class of symmetries that are directly connected with dynamics, the *gauge symmetries*.

5.1   The Master of the Gauge

Gauge symmetries have two roots: one that is rather pragmatic, stemming from electrodynamics and quantum mechanics, and a deeper one that comes from general relativity and is forever connected with the name of Hermann Weyl.

In the electrodynamics of the nineteenth century, the electric and magnetic potentials played a major role. When attempts were made to explain electrodynamic phenomena using mechanical models of the ether, one tried to deduce these potentials from the ether's supposed properties. Apart from the electric potential, which is familiar to us because a voltage simply represents a difference in this potential, a magnetic potential was also introduced, called the *vector potential*.

From these potentials, the electric and magnetic fields can be easily calculated. The fields, in turn, are directly measurable through the forces they exert on charged bodies. It would take us too far afield to go into the history of electrodynamics; I will content myself here with a short description of the state of affairs at the beginning of the twentieth century. In fact, it turned out that the decisive quantities were not the potentials, but the electric and magnetic fields themselves, the fields that appear in the Maxwell equations—the fundamental equations that govern electrodynamics.

The potentials, although quite useful for solving the equations, seemed to be only auxiliary quantities. The electric potential and the three components of the magnetic vector potential can be combined to form a four-vector, which has simple properties under the transformations of special relativity. We will refer to this four-vector as the *electromagnetic potential*. However, the electromagnetic potential is not uniquely determined: different potentials can lead to the same electromagnetic fields, and thus to the same physical behavior. A change of the potential that does not change the electromagnetic field is today called a *gauge transformation* or a change of gauge. The resulting symmetry actually seems rather banal—a gauge transformation of the potentials changes only quantities that are not directly observable, while the fields that are the observable quantities remain unchanged.

Nevertheless, these potentials play an important role in the formal treatment of classical mechanics. If one attempts to incorporate electromagnetic forces into the more refined formalism of advanced mechanics, it is in fact necessary to insert the potentials into the fundamental expressions, and not the fields. This was an important feature in the transition from classical to quantum mechanics, as we shall soon see.

Before discussing quantum mechanics, I would like to discuss briefly the second root of gauge symmetry, namely, general relativity. The starting point of general relativity is a principle stating that all coordinate systems are equivalent. There is nothing like a fundamental coordinate system that is privileged over all other reference frames. Therefore, the directions of two vectors (that is, quantities such as the velocity, which have both a magnitude and a direction) at two different points $\vec{P}_1$ and $\vec{P}_2$ cannot be directly compared.

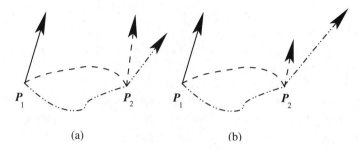

(a) (b)

Figure 5.1. (a) Parallel transport of a vector according to Einstein's theory of general relativity and (b) according to Weyl's near-geometry. Einstein's starting point is Riemannian geometry, where the direction may change along the path, but the length remains unchanged. In Weyl's near-geometry, however, the length can depend on the path as well.

To make a comparison, we must *parallel transport* the vector at point $\vec{P}_1$ to point $\vec{P}_2$, or the vector at $\vec{P}_2$ to $\vec{P}_1$. In Euclidean geometry, which is the geometry with which we are familiar and that governs our intuition, this is very simple, but more generally, the parallel transport of vectors will depend on the choice of the path taken. If a vector is transported by two different paths to the same point, the final direction will generally be different, as shown in Figure 5.1(a).

The difference between the two final directions is a measure of the curvature of space, and in general relativity it is the curvature that determines gravity, and vice versa. The length of a vector, however, remains unchanged. This is known as Riemannian geometry, which can be described as having a rigid gauge.

Weyl went one step further in these considerations. He assumed that in addition to the direction, the length of the vector could change during transport as well; see Figure 5.1(b). He viewed this approach, which he called *near-geometry* (*Nahgeometrie*, geometry in the small) as a natural extension of Riemannian geometry and thus of the theory of general relativity. Weyl did not mean that total arbitrariness prevailed in gauging the length at different points. Indeed, a path-dependent modification of the length must have a physical cause. In general relativity, the path-dependent change in the direction of vectors is determined by the gravitational field. One must know the solution for this field in order to parallel transport a vector. Conversely, the mathematical formalism of parallel transport incorporates a field that can be interpreted as the gravitational field. This is the famous geometric interpretation of gravity given by general relativity.

In the case of path-dependent lengths, Weyl was able to show that an additional field needed to be introduced. This field, required by Weyl's near-geometry, had precisely the properties of the electromagnetic field. Weyl thus believed that with this extension of general relativity he had found a unified geometric theory of both gravity and electrodynamics.

Weyl's theory was generally considered to be ingenious, but it met with skepticism with regard to its physical relevance. Weyl was slowly convinced by his critics, including Einstein, and later wrote:

> My first attempt to develop a unified theory of gravitation and electromagnetism dates from 1918, based on the principle of gauge invariance, which I had put alongside that of coordinate invariance. I have long since given up on this theory, after its correct core, that of gauge invariance, was appropriated by quantum theory as a principle that relates not gravitation, but rather the wave field of the electron, to the electromagnetic field.

Because of this original application to changes in the length, however, the names "change of gauge," "gauge transformation," and "gauge symmetry" remain.

In this remark, Weyl mentions the key words "quantum theory." Erwin Schrödinger was motivated by "Weyl's world geometry" to write a paper called "On a Remarkable Property of the Quantum-Orbits of a Single Electron." This work was based on Bohr's old atomic model, before the invention of the Schrödinger equation. Schrödinger was unable to draw a clear conclusion in this paper, but it was stimulating, and its significance became clear after the discovery of the Schrödinger equation.

As mentioned above, it is the potential, and not the fields themselves, that appears in the fundamental expressions of the advanced formulation of mechanics. This looks at first like a violation of gauge invariance, since we know that different potentials may lead to the same fields, and hence to the same observable results. In classical mechanics, this riddle is solved by the fact that in the equations of motion, which are derived from the initial fundamental expressions, only the fields appear, in a seemingly miraculous way.

In quantum theory, however, the situation is not so simple. Here the electromagnetic potential appears directly in the Schrödinger equation, and we are confronted with the following problem: what happens if we change the gauge, that is, if we change the potential in such a way that the fields themselves remain unchanged? This question was answered in 1926, the same year that Schrödinger published his equation. Vladimir Fock, from Leningrad (now once again Saint Petersburg), wrote a paper entitled "On the Invariant Form of the Wave and Motion Equations for a Charged Point-Mass." In this paper, Fock showed that the wave equation does not change under gauge transformations, as long as the wave function of the charged particle is modified together with the electromagnetic potential. Thus, observable results are not influenced by these transformations. Fock quoted Schrödinger's earlier paper, but he did not discuss Weyl's theory. However, the great significance of Weyl's theory was immediately recognized by Fritz London, who had independently come to the same conclusions as Fock. The title of London's paper, published in 1927, is "Quantum-Mechanical Interpretation of Weyl's Theory."

I will not discuss this important work, but will instead come directly to Weyl's decisive publication of 1929. To do so I shall have to return to the essence of gauge invariance. As an analogy, I shall discuss the introduction of a new currency, the euro, in several European countries. First, note that a global transformation is a simple matter; nothing changes, only the names. Looking at just a single country, the change from old to new currency is an example of such a global transformation. In Germany, one had simply to divide the *Deutschmark* prices by 1.9558 to obtain the new prices in euros. The buying power of one's income was uninfluenced by this change (or at least should have been). However, what for one country looked like a "global transformation" turned out to be a local transformation when seen as a whole. The exchange factor of 1.9558 is of little use to an Italian who wants to convert liras to euros; he needs a different conversion factor, valid for Italy. For countries outside of the euro zone the conversion factor depends not only on space (the country), but also on time, since it follows the course of varying exchange rates. If we want to compare prices before and after the introduction of the new currency, we need a conversion table. Only once we have checked with this table that real prices have not changed can we speak of a "change of gauge." Otherwise, we find that there is a real change in the price (a physical change, so to speak).

In Weyl's old theory, the change in length was a gauge transformation of this type. In his new theory, however, it was the change in a certain property of the charged particle's quantum-mechanical wave function. In both cases, the conversion table could be calculated from the change in the electromagnetic potential. One can, however, reverse the roles, and Weyl did exactly that: if we demand that gauge invariance hold, there must exist a conversion table, and hence an electromagnetic potential, that gives rise to the electromagnetic field.

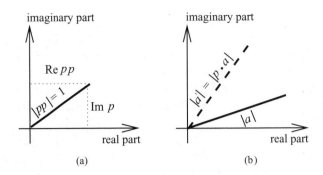

Figure 5.2. Geometrical representation of a phase factor and its action on a complex number in the complex plane (also called the Gaussian plane of numbers). (a) For a phase factor p, the modulus $|p|$ is equal to $\sqrt{(\mathrm{Re}\,p)^2 + (\mathrm{Im}\,p)^2} = 1$. (b) The multiplication of a complex number a by a phase factor p does not change its modulus: $|a'| = |p \cdot a| = |p| \cdot |a| = |a|$. It does, however, change its direction (phase).

As mentioned above, the postulate of gauge invariance of lengths was mathematically consistent, but was apparently not realized physically. However, this generalized type of gauge invariance was realized in quantum mechanics, and in fact turned out to be physically relevant there. Global invariance follows from the basic principles of quantum mechanics alone. Since the probability is given by the square of the modulus of the probability amplitude, any change in the wave function that does not affect this modulus will not change the observable quantities.

I already have mentioned this rather counterintuitive but essential feature of quantum mechanics in Section 3.2. This can be formulated in the following mathematically concise way: if the wave function of a state is multiplied by a complex number of modulus one, the modulus of the wave function is not changed, and the results of any observation made on this state do not change. Such a modulus-one complex number is called a *phase factor*. A geometrical representation of the phase factor and its properties is displayed in Figure 5.2.

Wave functions are extended objects, and in principle there is no limit on their extent. Weyl felt that it was unreasonable to demand that the phase factor introduced above take the same value everywhere, and he rejected what he called "parallelism at a distance." He therefore postulated that the phase factor can vary arbitrarily in space and time. Of course, it must remain a phase factor—a complex number of modulus one that does not contribute to the squared modulus of the wave function. If the phase factor can vary in this way,

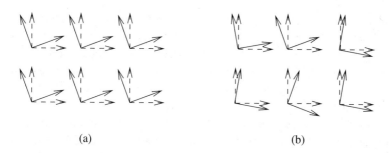

(a)   (b)

Figure 5.3. (a) Global rotation in the plane. In a global rotation, the rotation angle is the same at every point. (b) Local (gauged) rotation in the plane. In a local (gauged) rotation, the rotation angle can be different at different points.

then there is a local gauge symmetry, and one needs a conversion table to relate physically equivalent wave functions. The existence of such a conversion table in turn implies the existence of an electromagnetic field. Thus, in quantum physics the postulate of local gauge invariance requires the existence of electromagnetic interactions!

Let us summarize and introduce a few technical terms that will be useful for the following discussion. A transformation that is independent of space and time is called global. One then speaks of gauging a global transformation when one allows the transformation to act differently at different space-time points. Such transformations are called (local) gauge transformations. Figure 5.3 (a) shows a global rotation and Figure 5.3 (b) shows a gauged rotation.

A theory is called *gauge-invariant* if the fundamental equations are invariant under a gauged transformation. As we have seen, the requirement of gauge invariance has dynamic consequences. In order to implement a conversion table between physically equivalent configurations (which differ by a gauge transformation), one needs to introduce *gauge fields*. These fields compensate for the gauge transformations of the original fields in the fundamental equations, allowing these equations to remain invariant.

In practice, the procedure is as follows. If we begin with a theory of free (that is, noninteracting) electrons, the basic equation (in this case the Dirac equation) is invariant if the wave function of the electron is multiplied by a constant phase factor. Further, one can show that this global symmetry implies charge conservation for the electron. If the phase factor is gauged, however, such that it depends on space and time, the equation is no longer invariant. In order to restore invariance to the equation we must modify it through the introduction of gauge potentials. This must be done in such a way that under the *combined* transformation of the original wave function and the gauge potentials the equation keeps its original form. In this case the gauge potentials are precisely the electromagnetic potentials. Not only is their existence required by the principle of gauge invariance, but also the form of interaction with the electron is determined by the gauge symmetry.

Multiplication by a phase factor (a complex number of modulus one) is a particularly simple transformation group. To prepare the way for some of the jargon involved with more complicated transformations, we note the following. A number can be considered as a one-dimensional matrix, that is, a 1×1 table, with one column and one row. Therefore, the group of transformations consisting of multiplication by a phase factor is called U(1). The 1 shows that the dimension is 1, and U (which stands for "unitary") means that the number has modulus one. The corresponding symmetry is thus called U(1) symmetry. We therefore say that the electromagnetic field is the gauge field of the U(1) symmetry of charged particles.

In the latter half of the twentieth century, the principle of gauge invariance turned out to be the decisive tool for the construction of new and extremely successful theories. As an aside, we note that the mathematical framework needed for general gauge theories is the theory of fiber bundles. In this case the field is the fiber as a whole, while different potentials that yield the same physical field correspond to points along the fiber.

5.2   New Dimensions for the Gauge

It is not too far-fetched to take the next step, to gauging symmetries that are more complex than U(1), such as the groups SU(2) and SU(3). Attempts in this direction were made by Oskar Klein (1938) and Wolfgang Pauli (1953). Klein was stimulated by earlier work of Yukawa and came to a theory that would today be called an (explicitly) broken SU(2) gauge theory.

As for Pauli, who was very often critical of Weyl's attempts, he was interested in this problem for a short time. He wondered whether it was possible to gauge the SU(2) (isospin) symmetry in the same way as the U(1) symmetry. He succeeded in constructing an analogue of the electromagnetic field, but he did not pursue his investigation any further; presumably, he saw the difficulties involved in incorporating quantum physics. The problem was subsequently solved on the purely classical level, that is, without addressing the problem of quantization, by C. N. Yang and R. L. Mills in 1954, and independently by R. Shaw in his doctoral dissertation of 1955.

In gauging the U(1) symmetry, one has to introduce a gauge potential, from which the gauge fields—in this case the electric and magnetic fields—can be derived. The groups SU(2) and SU(3) are more complicated, and more generators (3 and 8, respectively) are needed to construct them. Therefore the group SU(2) (SU(3)) needs three (eight) gauge potentials.

The gauge fields for these symmetry groups are similar to those for U(1) in that they are derived from the gauge potentials and they have analogues of both electric and magnetic fields. Therefore, like the photons of electromagnetism, the field quanta of the gauge fields have spin 1 and are massless. These quanta are called *gauge bosons*. The interactions of gauge bosons are completely determined by gauge symmetry, and the only free parameter is the strength of the interaction. There are as many gauge potentials as there are generators for the gauge group. Since the group U(1) has only one generator, there is only one gauge potential, the electromagnetic potential.

The gauge symmetry group of electrodynamics, U(1), is especially simple, since the product of two phase factors is independent of their order. The result of the gauge transformation is therefore independent of the order in which the transformations are performed. It does not matter if I first multiply by p and then by q or first by q and then by p. As we learned in Section 1.5.1, groups with this property are called abelian. However, for the groups SU(2) and SU(3) the result generally depends on the order of multiplication. The same is true in general for any group SU(n), where n is an arbitrary integer greater than one. Such groups are said to be nonabelian (see also Section 1.5.1). One therefore says that electrodynamics is an abelian gauge theory. Theories with gauge invariance under nonabelian groups like SU(2) and SU(3) are called nonabelian gauge theories or Yang–Mills theories.

The field equations for electromagnetics (the Maxwell equations) are linear: they contain no terms that are of quadratic or higher order in the fields. Not only is this very convenient for their mathematical treatment, but it also has important physical consequences. The photons, which are the quanta for the gauge field (the gauge bosons), do not directly interact with each other. They are electrically neutral. In the language of Feynman graphs this means that there is no direct coupling of three or more photons.

The case for nonabelian gauge theories is different. Here, the field equations are no longer linear, and they contain terms that are both quadratic and

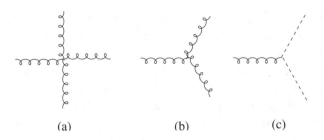

(a)   (b)   (c)

Figure 5.4. Interactions that occur only in a nonabelian gauge theory. (a) Interaction of four gauge bosons. (b) Interaction of three gauge bosons. (c) Gauge boson–ghost interaction. The wavy lines designate gauge fields and the dashed lines are ghost fields; the latter occur only as inner lines.

cubic in the fields. This reflects the fact that the gauge field quanta couple directly to each other. Another way of stating this is that in a nonabelian gauge theory the gauge bosons carry a charge (which is, however, different from the electromagnetic charge). Figures 5.4(a) and (b) show the Feynman graphs for interactions among nonabelian gauge bosons.

The coupling of nonabelian gauge bosons with each other makes the incorporation of quantum physics quite difficult. The path to a consistent *quantized* nonabelian gauge theory was "long and painful," as M. Veltman remarked. Of course, there was no hope that one would be able to do better than in quantum electrodynamics, so an attempt was made to establish a perturbation expansion analogous to QED (see Section 2.4).

In transposing the rules of QED to nonabelian gauge theories, incorporating the interaction of gauge bosons among themselves through the addition of interaction vertices (Figures 5.4(a) and (b)) to the usual interaction with fermions (which is the same as Figure 1.10), one is met with a nasty surprise: the scattering amplitudes obtained with these graphs violate unitarity, which in turn implies that probability is not conserved. After the era of nuclear democracy, physicists were very sensitive to this problem, and the inconsistency had to be corrected.

This was achieved through the addition of new fields that do not occur at all in the classical theory. Such fields, which appear only virtually (as interior lines in Feynman graphs), are called *ghost fields*. The interaction of ghost fields with gauge bosons is shown in Figure 5.4(c). It is only with these ghost fields taken into account that nonabelian gauge theory can be quantized in a way that respects special relativity. The technical difficulties are enormous and cannot be conveyed here, so I will only note the results of the final theory, which was established around 1971.

At that time, nonabelian gauge theories were found to be renormalizable, meaning that one could in principle calculate, with a limited number of parameters, up to any order of perturbation theory, and therefore make very precise predictions. Many physicists contributed to the quantization of nonabelian gauge theories, and the process was completed by 't-Hooft and Veltman.

Gauge invariance plays an essential role in renormalizability. Without gauge invariance, one cannot rule out the occurrence of more and more new terms, with undetermined couplings and thus additional parameters to be supplied by hand, at higher orders in the perturbative expansion. In particular, gauge invariance forbids the direct appearance of a mass term for gauge bosons, which would in turn destroy renormalizability. In QED this can be seen quite easily, but not in a nonabelian theory. In fact, Yang reports that during a seminar he was giving in Princeton in 1954 he was pressed so hard by Pauli regarding the mass of the gauge boson that he interrupted his talk. He continued the talk only after the intervention of Oppenheimer, the seminar's host.

Nonabelian gauge theories met with considerable interest in the early 1960s, before the questions of quantization and renormalization were settled. After the discovery of the octet of vector mesons, it seemed likely that these were gauge bosons of flavor $SU(3)$. It was therefore important to solve the problem with the mass, since all of these vector mesons had masses greater than $700 \, \text{MeV}/c^2$. On the other hand, the vanishing mass of gauge bosons seemed to be an essential feature of the theory; the gauge symmetry thus had to be broken in some way. In the theory of strong interactions these considerations brought no progress, but for the weak interactions they led to a breakthrough that began in 1967 with a paper of S. Weinberg. To understand this breakthrough, we have to learn another important concept, known as *spontaneous symmetry breaking*.

5.3   Spontaneous Symmetry Breaking

The distinction between a broken symmetry and no symmetry at all is rather subtle, like that between a badly preserved castle and a well-preserved ruin. Therefore, a broken symmetry is always somewhat suspect. There is, however, a very respectable and well-defined way of breaking symmetry, known as spontaneous symmetry breaking, which we will deal with in this section.

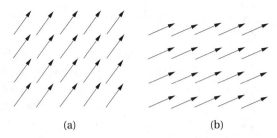

(a)   (b)

Figure 5.5. An iron magnet (or ferromagnet) is a system with infinitely many ground states. The energy is minimized if all spins, here indicated by arrows, point in the same direction. The direction in which the spins actually point, as shown in (a) or (b), is irrelevant for the energy.

In the quantum physics of fields there is a privileged state, the ground or *vacuum* state, which is the state of lowest energy. From it, all other states can be constructed through the application of field operators. This feature is even incorporated in the axiomatic formulation of quantum field theory, defined by the Wightman axioms; see Section 2.4. A symmetry is said to be *spontaneously* broken if the fundamental equations that describe the theory are invariant under the symmetry, but the minimal-energy ground state is not.

Spontaneous symmetry breaking can be most easily illustrated with an example from solid-state physics. In this case, the concept was first introduced by Heisenberg in 1928. Let us consider a magnet consisting of atoms with some particular spin (known as a Heisenberg magnet). Spin is a directed quantity: the interaction of two neighboring atoms depends only on the angle between the two spin directions. If the two spins are parallel, the interaction energy is minimal. This spin–spin interaction is invariant under (global) rotations, since such a global rotation will not change the difference in angles between two spins. The total energy is minimal if all spins are parallel. Therefore, in the ground state all spins point in the same direction. However, it does not matter in which common direction they point. Therefore, there are in fact infinitely many ground states, one for every possible direction. Figure 5.5 shows two possible ground states.

In such a case one speaks of a *degenerate ground state* or a *degenerate vacuum*. For real magnets, the spins must point in some direction. This direction is thus privileged, and the original rotational symmetry is called *spontaneously broken*.

The degeneracy of the ground state has one far-reaching consequence: since no energy is needed to make the transition from one ground state to another, there are massless particles (in solid-state physics these are called quasi-particles) that induce transitions between different ground states. This is a very important general consequence of spontaneous symmetry breaking: if a continuous symmetry is spontaneously broken, there must be massless particles. This theorem was established by J. Goldstone in 1961. The massless particles, which further must have integer spin, are called *Goldstone bosons*. The Goldstone bosons for a ferromagnet are called magnons, the field quanta of spin waves. Perhaps better known are the field quanta of sound waves, called phonons, which arise as Goldstone bosons from the spontaneous symmetry breaking of translation invariance.

In elementary particle physics, the concept of the ground state is not as intuitive as in solid-state physics, but the mathematical treatment is the same. In particle physics the ground state is generally called the vacuum state or simply the vacuum. This state plays an important role in quantum field theory, which can be seen from the Wightman axioms (see Section 2.4). The first proposal to apply the concept of spontaneous symmetry breaking in particle physics also goes back to Heisenberg, in 1959. However, Heisenberg's nonlinear spinor theory, to which the idea was applied, was not very popular, and only the papers of Y. Nambu, which appeared roughly two years later, were taken note of.

We will investigate spontaneous symmetry breaking through a simple but very important field-theoretic example. We consider (provisionally) a particle without spin and without charge. Such a particle is described by a real field F. The field F depends on space and time, but for the moment this is not important. The field interacts with itself, and the field quanta have mass m. The mass m contributes to the field energy through the quadratic term $m^2 \cdot F^2$, corresponding to the rest energy of the field quanta. The self-interaction yields a contribution with a fourth power of the field, $g \cdot F^4$, where the positive coupling g describes the strength of the interaction. The total static field energy is thus given by

$$V(F) = M \cdot F^2 + g \cdot F^4,$$

where $M = m^2$.

The static field energy V as a function of the field strength F is displayed in Figure 5.6(a). One can see that it is symmetric under reflection around the V-axis, meaning that it is invariant under a change of sign of F. The ground state, which has minimal energy, is at $F = 0$. If instead we choose to ignore the fact that M is the square of a mass and let it be negative, the situation is quite different. This sounds a bit strange—and indeed it is—but let us see what we

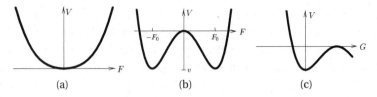

Figure 5.6. (a) Static field energy with a positive quadratic term (a standard mass term). (b) Static field energy with a negative quadratic term, leading to spontaneous symmetry breaking. (c) Static field energy where the field strength has been shifted by the value F_0: $G \quad F \quad F_0$.

can learn from the formalism. If it behaves reasonably, we should be able to interpret the result reasonably.

At small values of F the term $M \cdot F^2$, which is now negative, dominates, and the field energy is negative. At larger values of F, however, the interaction term $g \cdot F^4$ is dominant; we obtain the curve displayed in Figure 5.6(b). Through the interplay between the "repulsive" quadratic term and the "attractive" interaction term, the state of minimal energy is no longer located at $F \quad 0$, but at the finite values F_0 and $-F_0$, where $F_0 \quad \sqrt{-M/(2 \cdot g)}$ (please note that $-M$ is positive here). We now have two possible ground states (or vacua), which have nonzero field strengths F_0 and $-F_0$. The field energy V has the same value in either vacuum. The minimal value taken by the field energy is $v \quad -M^2/(4 \cdot g)$. Now, in order to quantize the system, one has to start from a definite ground state (vacuum). Therefore, depending on which of the two vacua we quantize around, the symmetry under the sign change of F is spontaneously broken. Recall that a symmetry is said to be spontaneously broken if it is valid for the underlying fundamental equations (as it is here for the potential field energy), but not for the specific ground state chosen by the system.

Our next step is to introduce a new field, namely $G \quad F \quad F_0$. The new field G simply represents a shift of the field F by the constant F_0; the field energy as a function of the new field G is shown in Figure 5.6(c). It is obtained from Figure 5.6(b) by shifting the V-axis to the left. The formula for this curve is easily obtained by replacing the variable F in the above expression for V by the variable $G - F_0$:

$$V \quad -2M \cdot G^2 - 4g \cdot \sqrt{\frac{-M}{2 \cdot g}} \cdot G^3 \quad g \cdot G^4 - \frac{M^2}{4 \cdot g}.$$

The expression now looks more complicated; in particular, there is now a term with G^3, a cubic term that destroys the symmetry under the transformation of F into $-F$. This is a consequence of spontaneous symmetry breaking; we have decided to use $G = F + F_0$, but we could just have well used $G = F - F_0$. Though this expression for V looks more complicated, it has a big advantage: the minimum of the field energy V now occurs for a field strength $G = 0$. This is reflected in the *positive* quadratic term $-2M \cdot G^2$, which now allows for the following interpretation of the model: it describes a particle with mass squared equal to $-2M$, and with cubic (G^3) and quartic (G^4) self-interactions. Note that we can view the model with two different appearances: for symmetry considerations it is convenient to work with the field strength F, while for the particle interpretation of a quantum field theory it is more convenient to work with G.

In order to make this model truly interesting for field theory, we have to extend it. The symmetry operation "change the sign of F" is not a continuous operation, and in fact we have seen that Goldstone bosons occur only if a *continuous* symmetry is spontaneously broken. Therefore, we will rotate the figures from Figure 5.6(a) and (b) around the V-axis. From the curve in Figure 5.6(a) (with positive M) we obtain a bowl, Figure 5.7(a), while from the curve in Figure 5.6(b) we obtain a surface that is similar to the bottom of a champagne bottle, Figure 5.7(b). Expressed in mathematical formulas, rotating the curves in Figure 5.6 corresponds to the introduction of two fields, F_1 and F_2, with a static field energy given by

$$V(F_1, F_2) = M \cdot \left(F_1^2 + F_2^2\right) + \lambda \cdot \left(F_1^2 + F_2^2\right)^2.$$

For negative M we now have a continuously infinite number of ground states, namely the groove in the bottom of the champagne bottle. The field strengths for these ground states satisfy the equation

$$\left(F_1^2 + F_2^2\right) = -\frac{M}{2 \cdot \lambda}.$$

The field energy, as well as the corresponding dynamics, is invariant under all transformations that leave $F_1^2 + F_2^2$ unchanged. This corresponds exactly to a rotation in the F_1–F_2 plane. In formulas, this rotation is expressed by the substitution

$$F_1 \to \cos\theta \cdot F_1 + \sin\theta \cdot F_2; \quad F_2 \to -\sin\theta \cdot F_1 + \cos\theta \cdot F_2,$$

where the field strengths F_1 and F_2 depend on space and time, but the rotation angle θ is constant (in other words, we are considering a global transformation of the fields).

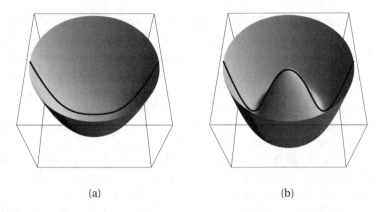

(a)   (b)

Figure 5.7. (a) Static field energy $V(F_1, F_2)$ of a field with two components F_1 and F_2 with positive quadratic term (mass term). (b) Static field energy $V(F_1, F_2)$ of a field with two components F_1 and F_2 with negative quadratic term, leading to spontaneous symmetry breaking.

It is very convenient to combine the two fields F_1 and F_2 into a complex field, where now F_1 and F_2 are respectively the real and imaginary parts. Now the symmetry operation, rotation in the F_1, F_2 plane, corresponds to multiplication by a phase factor, meaning that this is a U(1) transformation. We can summarize the situation as follows: the model with negative M exhibits a spontaneously broken U(1) symmetry. It has infinitely many ground states, but if we want to describe a real system, we have to choose one in particular, say $F_{10} = F_0$, $F_{20} = 0$, and the symmetry is broken by this particular choice.

Since in this case the symmetry is continuous, Goldstone's theorem applies, and there will be a field with massless quanta. This can be seen by substituting the fields F_1 and F_2 by new ones, as we have done previously in the case of a single field. The calculation is not particularly complicated. Along with the field with massless particles, there is also a field with (positive) mass squared equal to $-2M$.

Spontaneous symmetry breaking brings to mind the paradox of Buridan's ass, which was dear to medieval scholastics. This ass was said to have starved to death because, being a completely rational donkey and finding himself positioned exactly between two identical haystacks, he could not choose between them. It is my guess, however, that in reality the donkey was well aware of the possibility of spontaneous symmetry breaking, and spontaneously decided on one of them.

5.4   The Higgs–Kibble Dinner

Let us go one step further. The rotation in the plane of the two fields F_1 and F_2, introduced in the last section, was a global symmetry transformation, since the angle θ was independent of space and time. If Weyl's gauge principle is applied to this symmetry, the angle θ can now depend on the space and time coordinates. In order to preserve the gauge symmetry, one has to introduce gauge fields. We have learned that application of the gauge principle to (unbroken) U(1) symmetry leads to electrodynamics and the existence of photons.

As early as 1938, Oskar Klein tried to explain strong and weak interactions by the exchange of gauge bosons. Since charge is exchanged in these interactions, these bosons have to carry charge. Klein worked with SU(2) as the transformation group. His work would have perhaps had more influence if the interests of physicists in the following years had not been directed in completely different directions. In 1958, S. Bludman proposed gauge bosons as the transmitters of weak interactions. Since he used SU(2) as the symmetry group, he concluded that there must also be a neutral gauge boson, which mediates so-called weak neutral current interactions. This conclusion was derived along the same lines as those of Kemmer on the existence of neutral pi mesons (Section 1.5.3). From the beginning, however, it was clear that it would be extremely difficult to detect these neutral currents. We will return to them in detail in Section 6.3.

In fact, the masses of the gauge bosons were a much bigger problem than the yet-to-be-discovered neutral gauge boson. It had been known since the early 1930s that weak interactions are very short-ranged, and therefore that the particles that cause them had to be very heavy. This was extremely awkward for gauge theories, since gauge theories predict massless gauge bosons. Therefore, the gauge symmetry had to be broken somehow. It was supposed—as it turned out, rightly so—that explicit symmetry breaking (where the fundamental equations themselves do not respect the symmetry) would not allow one to construct a renormalizable theory of weak interactions. Therefore, the symmetry breaking should be spontaneous. However, spontaneous symmetry breaking as we have described it seemed to be excluded, since it predicts massless Goldstone bosons, and these were not observed. However, in more detailed investigations of broken gauge symmetries, it turned out that once again a miracle happens, and two exact theorems compensate each other. This is the so-called Higgs–Kibble dinner (also called the Higgs–Kibble mechanism): on the occasion of spontaneous symmetry breaking, the gauge bosons "eat" the

Goldstone bosons and grow fat from this dinner, meaning that the Goldstone bosons disappear and the gauge bosons acquire a mass.

This drastically hand-waving description is perhaps unsatisfactory, so I shall give a somewhat more detailed description in words; the website for this book contains the mathematical details. We return to our fields F_1 and F_2 from the previous section and apply the gauge principle to the symmetry transformation of the fields, in this case the rotation around an angle θ in the F_1–F_2 plane. The angle θ depends on the space and time coordinates, and additional gauge fields are introduced to ensure invariance under local gauge transformations. The form of the interaction between the gauge fields and the original fields F_1, F_2 is fixed by the symmetry requirements, and therefore the static field energy depends on the gauge fields in a well-defined way. The symmetry is now spontaneously broken by fixing the ground state, say to $F_1 = -F_0$, $F_2 = 0$. After fixing the ground state, it is convenient to introduce new fields G_1, G_2, whose field strengths take the value zero for this ground state (as in Figure 5.6(c)), that is,

$$G_1 = F_1 + F_0, \quad G_2 = F_2 .$$

If the static field energy, which now also contains gauge fields, is expressed in terms of the new fields G_1, G_2, one obtains a quadratic term for the gauge potential, indicating that the gauge bosons have acquired a mass. The field for the Goldstone bosons is still present in the formulas, but there is no observable particle that corresponds to it. This can be seen from the fact that it can be "transformed away" by choosing an appropriate gauge. This is the previously mentioned Higgs–Kibble dinner: the Goldstone boson has been eaten by choosing a suitable gauge, and the gauge bosons are now massive. There is still a field quantum for the field G_1: it is a boson with spin 0 and positive mass squared $m^2 = -2M$. This particle is the so-called *Higgs boson*, which is now the most eagerly sought-after particle in physics.

This whole picture is rather bold, and it is not astonishing that in 1964, these considerations, discussed independently by P. W. Higgs and T. Kibble, as well as by F. Englert and R. Brout, met with skepticism. Higgs recollects that just before a seminar at Princeton, a rigorous field theoretician called his attention to the fact that he must be wrong, since the Goldstone theorem had just been proven within the mathematically rigorous framework of C^* algebra. Nevertheless, Higgs had not made an error, since in some respect, two rigorous theorems cancel each other out, namely that gauge fields are massless and that in a theory with a spontaneously broken continuous symmetry, Goldstone bosons appear.

The proof of the Higgs–Kibble dinner seems at first sight somewhat suspicious, since the vanishing of the Goldstone boson becomes apparent only in a special gauge, the so-called *unitary gauge*. If another gauge is chosen, is it then possible to work with massless gauge bosons and Goldstone bosons? To a certain extent the answer is yes: for intermediate calculations it might be convenient to work in a gauge different from the unitary one. For example, this freedom in the choice of gauge is very important for the proof of renormalizability.

In solid-state physics, the phenomenon of spontaneous gauge-symmetry breaking was first discussed in 1958 by P. W. Anderson in his investigations of superconductivity. Mathematical physicists point out that a gauge theory cannot be spontaneously broken if one does not confine oneself to perturbation theory, and what looks like a breaking is only a gauge fixing.

In more complex gauge theories, such as gauged $SU(2)$ and $SU(3)$, the results are fundamentally the same. In the final analysis, there are as many massive gauge bosons as there are generators of the group, i.e., three for $SU(2)$, and there are no longer any Goldstone bosons, but there are one or more massive Higgs bosons.

The triumphant end of all this theoretical development was the proof that spontaneous symmetry breaking leads to a renormalizable theory. This was completed in 1971 by G. 't Hooft, who had just completed his doctorate, and his advisor M. Veltman. As mentioned earlier, such a proof is a highly technical procedure, and here I will mention only that gauge symmetry is an essential feature of the proof. Explicit breaking of the symmetry, for example, through explicit nonzero-mass terms for the gauge bosons, would lead to nonrenormalizable interactions. In principle, quantum corrections could also lead to a fatal symmetry breaking. We will briefly discuss this issue in the next section.

5.5   Anomalies

In this section, we will briefly discuss another kind of symmetry breaking, which occurs only in quantum field theories. In this kind of symmetry breaking, both the fundamental equations and the ground state are invariant under the relevant symmetry transformations, but there are quantum corrections that violate this symmetry. These quantum corrections are called *anomalies*. They are due to the creation and annihilation of virtual particles and are therefore not directly accessible to our intuition. They have the potential to play a

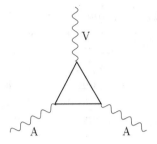

Figure 5.8. The triangular diagram representing violation of axial charge conservation. The external lines represent one vector field (V) and two axial vector fields (A). The inner lines represent massless fermions.

disastrous role in the weak interactions, but fortunately they do not do so, since once again a miracle happens, as we will see in Section 6.10. Therefore, I can be very brief and give only one example.

Let us consider a theory with massless quarks. On the classical level (that is, ignoring quantum corrections), we can construct two conserved charges within this theory. By "conserved" we mean that this quantity does not change with time. One of these charges can be interpreted as the electric charge, while the other is called the *axial charge*, which changes its sign under parity transformations. It turns out that quantum corrections lead to terms that violate the conservation of axial charge. The quantum correction responsible for this violation is the triangular graph displayed in Figure 5.8.

This graph plays an important role in the decay of the neutral pi meson into two photons. It was first discussed in 1949 by J. Steinberger, in his thesis on pi meson decay. However, Steinberger received the Nobel Prize not for this work, but for his experimental work in neutrino physics. The graph of Figure 5.8 was first recognized as the source of an anomaly in 1951 by Julian Schwinger, and was discussed extensively by S. Adler, J. S. Bell, and R. Jackiw in 1969.

5.6 Better Counters, Better Accelerators, and Better Beams

It would be a grave omission were we not to emphasize the decisive contributions of experimental physics on the way to the standard model. Theoretical developments did not take place in a vacuum, and in the beginning there was no satisfactory theory. The rise of the standard model was instead an evolutionary process, in which the effects of the environment, meaning the experimental results, played a decisive role. In this process, great progress in all fields of experimental particle physics was equally important, including progress in the construction of better accelerators, better detectors, and better particle beams.

An important contribution to the development of detectors came from solid-state physics, especially the physics of semiconductors. If a charged particle passes through a semiconductor, it does not knock electrons out of the constituent atoms, but only elevates them into higher-energy states, where they are able to contribute to current conduction. Much less energy is needed for elevation into a higher state than for complete ionization, and many electrons are available for the detection process, even in a small region. Semiconducting detectors are therefore of particular use where an extremely precise spatial resolution is essential, for example, if one wants to know where several particle trajectories meet. These are the so-called vertex detectors.

Another more indirect, but perhaps even more important, influence of semiconductor physics was the replacement of vacuum tubes by transistors. This opened up new possibilities in electronics, especially in computer construction. In bubble chambers, which were so important for the discoveries of the 1960s and 1970s, scanning had to be done by eye. The human scanners sat at large tables, looking for interesting combinations of tracks in the projected pictures. One can get an idea of the difficulties of this procedure if one looks at the picture of the first detected omega hyperon, shown in Figure 4.5.

When scanners found an important combination of tracks, they marked several points on the trajectories by hand, and these points were stored electronically. The rest was done by computer, which from the marked points reconstructed the particle trajectory in three-dimensional space and, using energy and momentum conservation, calculated the kinematic characteristics of the particles. Use of a Geiger counter was much more convenient, because the information could be registered automatically. However, the information obtained this way was meager: one did not know the position of the particle within the counter, and one was less able to reconstruct its trajectory.

Wire and spark chambers, developed from the Geiger counter, combine the advantages of counters and bubble chambers. A wire chamber has many wires, each of which acts as a counter. From information about which wire was excited, and at what time, the trajectory of the ionizing particle can be reconstructed as it was in the bubble chamber. The excitation of the wire can be easily registered electronically, whereas in the bubble chamber the points had to be marked by hand. Some of the programs developed for the reconstruction of trajectories in bubble chambers could be used directly for the analysis of data from wire chambers. There is, however, an enormous difference: in a bubble chamber experiment, only those trajectories that were seen as interesting were registered. In wire chambers, on the other hand, all trajectories are registered. This results in a gigantic flow of data. In order to limit the flow, a certain pre-selection of registered events is necessary. This leads to enormous technical problems and even—because of possible theoretical biases—to problems of a more fundamental nature. For his contributions to the development of wire chambers, G. Charpak was awarded the Nobel Prize in 1992.

In the preselection process, Cherenkov counters play an important role. These detectors emit light only if the passing particle has a velocity that is greater than that of light in the medium. Therefore, they respond only if the velocity is above a certain threshold. The threshold value can be fine-tuned, for example through varying the pressure in a counter filled with gas. The refraction index, and with it the velocity of light in the gas, depends on the pressure. Thus if one wants to register only very highly energetic particles, one registers an event only if a Cherenkov counter with a high threshold responds at the same time.

Hybrid detectors have gained particular prominence in more recent experiments. In these, many counters of different kinds are combined in order to detect particles with very different properties and energies. Figure 5.9 shows a photograph of such a detector. I cannot discuss the individual components, but the complexity of the apparatus is clearly visible.

A new type of accelerator was developed in the 1960s, the *collider*. In it, two particle beams are accelerated and brought to a head-on collision. By this procedure, the available energy is increased dramatically. In conventional accelerators, the highly energetic accelerated particles were directed onto a fixed target; most of the energy was expended in putting the resting target particle into motion, and only a small part was used, for example, for the production of new particles. The "useful" energy is the *center of mass* (or center of momentum) energy; it is the energy of the system if the two particles have equal momentum but opposite direction. Let us consider an example: if one wants to produce a particle of 90 GeV/c^2 in a proton–antiproton collision, and for that purpose directs highly energetic antiprotons onto protons at rest, in other words a fixed hydrogen target, the energy of the accelerated antiprotons must be at least 4,510 GeV. However, if the protons as well as the antiprotons are

Figure 5.9. The detector Mark I from the Stanford Linear Accelerator Center (SLAC), with Roy Schwitters in the interior.

accelerated and are brought to a head-on collision, each beam has to be accelerated to an energy of only 45 GeV.

If the particles in the two colliding beams have different charges, for example if they are particles and their associated antiparticles, the acceleration and storage of the two beams can happen in the same circular accelerator, since the directions of the orbits are opposite for particles of opposite charge. The beams are designed in such a way that they deviate slightly from circular orbits and intersect at certain points where the experiments are performed. After acceleration, the particles can be stored and "recycled" in the accelerator or in specially designed storage rings. Bruno Touschek rendered great service to the development of colliders and storage rings. Around 1964, in his laboratory in Frascati, Italy he put into operation the first storage ring containing both electrons and positrons. Although the available useful energies in colliders are much higher than in fixed-target machines, one has to pay the price that there is a much smaller number of events. Therefore, collider experiments normally run for a very long time.

Higher energies imply larger accelerator diameters, since the diameter of the particle's orbit in the magnetic field increases with energy. In order to stay within affordable limits for the dimension of the accelerator, one has to produce very strong magnetic fields. In recent times, superconducting magnets have begun to be used. These can produce very high magnetic fields, but they have to be cooled with liquid helium, which entails additional advanced technology.

For electrons, there is yet another effect that prohibits the use of small accelerator diameters. Orbiting charged particles emit electromagnetic radiation and thus lose energy. These radiation losses increase with energy, and are all the greater, the smaller the mass and the smaller the diameter. Therefore, circular high-energy electron accelerators have to be quite large. The large electron positron collider (LEP) at CERN has a diameter of more than five miles (8.5 km).

Therefore, one frequently uses linear accelerators for electrons. As indicated by the name, the electrons move along a linear trajectory; they surf on an electric traveling wave like a surfer on an ocean wave. Since electrons are not trained surfers, the design must be very refined, so that the electrons can follow the electric wave and not get lost.

In Figure 5.10, we show the development of available particle energies over the course of the twentieth century. Here, we have displayed the available (useful) energy, that is, the energy in the center-of-mass system. As can be seen from the dashed line, the available energy has grown by a factor of approximately ten every sixteen years.

There was also great progress in particle beam construction, made possible through very cleverly designed magnets. A beam of neutrinos was produced at the Brookhaven National Laboratory. They originated in the decay of pi and K mesons. The mesons were produced in large numbers by bombarding a target in the accelerator with accelerated protons focused by magnetic lenses. Since the decaying mesons have high velocity, the neutrinos move predominantly in the forward direction as well, and one obtains a well-focused neutrino beam.

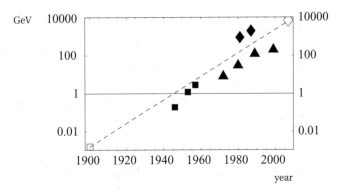

Figure 5.10. Maximal available particle energies since 1900. Squares represent accelerators with fixed target, triangles are colliders with electrons and positrons, and diamonds are colliders with protons and antiprotons. The open square indicates the energy available from natural radioactive sources, and the open diamond represents the Large Hadron Collider (LHC) under construction at CERN; it will have an energy of 7,000 GeV per beam. The dashed line indicates an increase of energy by a factor of 10 every 16 years.

By using either positive or negative mesons one can obtain either neutrinos or antineutrinos. A similar neutrino beam had been constructed at CERN.

In experiments with these beams, an important result was obtained: neutrinos produced through decays whose final state included muons (e.g., the decay of a pi meson into a muon and a neutrino) would in turn always pro-

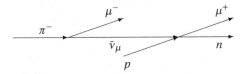

Figure 5.11. The reaction chain for muon production by (anti)neutrinos. The negative pi meson (π^-) decays into a negative muon (μ^-) and an antineutrino ($\bar{\nu}_\mu$). The antineutrino hits a proton (p) and produces a positive muon (μ^+), and the proton is transformed into a neutron (n). From the fact that neutrinos occurring in muonic decay modes create only muons, one can conclude that there is a special muon neutrino, different from the electron neutrino. Note that this is not a Feynman graph; the lines represent real particles.

duce muons, and never electrons or positrons. This showed that there are two kinds of neutrinos: the electron neutrino and the muon neutrino. The decay and production interactions of muon neutrinos are displayed in Figure 5.11.

The importance of these experiments was recognized by the awarding of the 1988 Nobel Prize for physics to L. M. Lederman, M. Schwartz, and J. Steinberger for "the neutrino beam method and the demonstration of the doublet structure of the leptons through the discovery of the muon neutrino." Experiments with neutrino beams made essential contributions to the further development of the standard model, as we will see in Section 6.3.

5.7   The Electron Microscopes of Particle Physics

Scattering experiments play a central role in particle physics. Particle spectroscopy started with the scattering experiments of Fermi and his collaborators; see Section 2.5. But even before this, scattering experiments were essential for investigating the structure of matter. We have seen, in Section 1.2, that Rutherford concluded from the scattering experiments of Geiger and Marsden that the atom has a tiny nucleus, within which practically all of its mass is concentrated.

However, there was an essential difference between the experiments of Geiger and those of Fermi. Geiger and Marsden investigated scattering of alpha particles that were the result of natural radioactive decay, and the deflection of the scattered particles was mainly caused by the electric field of the nucleus, that is, the electromagnetic interaction. In Fermi's experiment, however, the pi mesons from the accelerator interacted with the target protons mainly through the strong interaction. This difference comes about due to the different energies involved. In most cases, the alpha particles did not have enough energy to come sufficiently close to the nucleus to interact strongly; this was prohibited by the strength of the electromagnetic repulsion.

However, the pi mesons in Fermi's experiments had enough energy to overcome the electromagnetic repulsion and to penetrate the proton. At such small distances, the strong interaction dominates. Since the strong interactions are more complicated than electromagnetic interactions, it is difficult to obtain information on the structure of hadrons from high-energy hadron scattering experiments. Rutherford's conclusions on the structure of the atom, which he derived from scattering data, were convincing because he worked within the framework of the well-understood electromagnetic interaction. In fact, if one

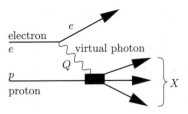

Figure 5.12. Electron–proton scattering. A highly energetic electron interacts with a proton and transfers a large amount of momentum Q. The exchanged virtual photon has a wavelength that is inversely proportional to the transferred momentum Q.

wants to perform high-energy experiments in which the electromagnetic interaction is the principal cause of scattering, one has to scatter particles that do not interact strongly. In this case, electrons are the simplest (and cheapest) choice.

One may ask why one wants to perform high-energy scattering experiments at all. The answer lies in the same physical principles that apply for the conventional microscope: if one wants to recognize small structures, one needs light whose wavelength is smaller than the size of those structures. By looking at the Feynman graphs introduced in Section 1.4.2 and the diagram in Figure 5.12, one can see the analogy between light microscopes and electron-scattering experiments graphically. If an electron scatters off of a proton, one can describe the process in quantum-field-theoretic language as follows: the electron emits a virtual photon, and with this photon one can "see" inside the proton.

In Figure 5.12, the virtual photon is represented by a wavy line, and the interaction with the proton by a black box. The wavelength λ of the virtual photon is inversely proportional to the momentum Q that it transfers from the electron to the proton, $\lambda = \hbar/Q$, where $\hbar$ is the Planck constant. In general, this interaction is inelastic, and the proton is transmuted into another hadronic state, consisting of either a resonance or a baryon and additional mesons; this final hadronic state is designated by X. The transferred momentum Q and the energy W of the hadronic state X will play an important role in the following.

The more energy, and thus more momentum, that an electron has before interacting, the more momentum it can transfer to the proton. High momentum transfer means a virtual photon with short wavelength, and thus good resolution. High energies are therefore necessary not only to produce heavy particles, but also in order to resolve fine structures. With a momentum transfer of 1 GeV/c, structures with a size of about 0.1 femtometer (10^{-16} meters) can be "seen." This is about one-tenth the size of a proton. At the Hadron Electron Ring Accelerator (HERA) (the first electron-proton collider in the world) located at the German Electron Synchrotron center (DESY), in Hamburg, a beam of electrons is directed against a beam of highly energetic protons. The maximal momentum transfer is here about 900 GeV/c, corresponding to a resolution of 0.0001 femtometers (10^{-19} meters).

The advantage of electron-scattering experiments, as opposed to pure hadron scattering, is that the electromagnetic interaction, in the framework of QED, is well understood. Therefore, reliable conclusions can be drawn from electron-scattering experiments about the structure of the target particles. Since there never is a free lunch, there is still a price to be paid. The electromagnetic interaction is much weaker than the strong interaction and therefore the cross sections are rather small, and in electron-scattering experiments there are fewer events than in experiments with hadron beams. Therefore, the experimental layout has to be well adapted to the question at hand, and in general the experiments take a long time.

In Figure 5.13, we show qualitatively how electron scattering can yield information on the structure of the target particle. The trajectories for a particle scattered off a pointlike target are shown in Figure 5.13(a); the solid curve represents a particle of high energy; the dashed curve, one of low energy. The higher the energy of the particle, the closer it can come to the target. The angular distribution, that is, the relative probability for scattering at a certain angle, is independent of the energy for a pointlike charge. For an extended charge distribution, the situation is different. A particle with high energy can penetrate deeply into the target, and therefore sees little charge between itself and the center of the target. Therefore it will be much less deflected by an extended charge than by a pointlike charge; see Figure 5.13(b). One also sees from the figure that only a highly energetic particle can penetrate the target deeply enough to test the charge distribution.

Though quantum physics complicates the results, classical considerations are still qualitatively valid in quantum field theory. It is customary to divide the observed angular distribution by the theoretical angular distribution for scattering from a pointlike scattering center; the result is called the *form factor*. The form factor depends in general on the momentum transfer Q. If the form factor decreases quickly with increasing Q, the target particle is an extended object; if it is constant, the target is pointlike. The form factor gives us something like an electron-microscope picture of the target particle. The form factors for the two cases under discussion are shown in Figures 5.13(c) and (d).

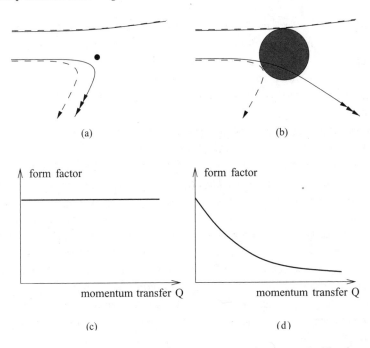

Figure 5.13. Trajectories of scattered particles. (a) Scattering on a pointlike charge distribution. (b) Scattering on an extended charge distribution. The dashed line applies to particles of low energy, the solid line to a particle with high energy. (c) and (d) The form factors for the two cases are qualitatively displayed; the form factor is the actual angular distribution of the scattered electrons divided by the theoretical distribution for scattering on a pointlike charge. Here the scattering of two charges with like signs is displayed; for the case of opposite charges, the same arguments hold qualitatively.

Geiger and Marsden observed an angular distribution that was very close to that of a pointlike scattering center; see Figure 1.4. The form factor is thus nearly constant. From this, Rutherford concluded that the atomic nucleus is nearly pointlike.

5.8 Deep Inelastic Scattering

In order to investigate the structure of the proton and the neutron, R. Hofstadter performed very precise scattering experiments with electrons of energy around 1 GeV. These experiments were done at the linear electron accelerator Mark III at Stanford University. He obtained the charge distribution through an analysis of the angular distribution of the scattered electrons. As expected qualitatively from meson field theory, the proton is indeed not a pointlike particle, but has a charge that is distributed in a cloud with a diameter of about one femtometer. The neutron has a charge distribution, where positively charged contributions near the center are compensated by negative charges in the outer regions.

Hofstadter's highly successful experiment—he was awarded the Nobel Prize in 1961—was certainly a motivation for the energetic pursuit of the program of high-energy scattering. In 1957, the Stanford group, headed by "Pief" Panofsky, submitted plans for a linear accelerator with a length of three kilometers. In 1961, after many political and scientific discussions, the plan was approved and supported with 114 million dollars. Panofsky was appointed head of the Stanford Linear Accelerator Center. For the first stage of the experiment, an energy of 20 GeV was planned, which was 20 times the energy of the precursor accelerator Mark III. The accelerator was built within the planned time and cost limits. Considering the immense technical and fundamental challenges encountered during construction, one cannot overestimate this achievement. The distinguished experimental physicist Panofsky thus also proved his extraordinary organizational abilities.

A group of physicists from Stanford University and the Massachusetts Institute of Technology continued the program initiated by Hofstadter at this new accelerator, but now the momentum transfer could be much higher. In their first investigations, they looked for elastic scattering, that is, processes in which the proton remains intact (the X in Figure 5.15 is still a proton). They confirmed the fast decay of the form factor with increasing momentum transfer. In 1967, they began the main program. They investigated processes in which the proton is transformed into an excited baryon resonance. Here again, the form factor decreased rapidly with momentum transfer. The big surprise came from events that were referred to as *deep inelastic scattering*, meaning that the momentum transfer is larger than around 1 GeV/c and that the process is highly inelastic, so that the total energy of the produced hadronic state X is more than double the rest energy of the proton (about 2 GeV). As can be seen from Figure 3.4, for such energies we are beyond the region where resonances are found.

For these deep inelastic events, in which the electron can penetrate "deeply" into the proton, there was an unexpected result: for large momentum transfer the form factor stays nearly constant. This is shown schematically in Figure 5.14. The form factors for deep inelastic scattering processes are called *structure functions*.

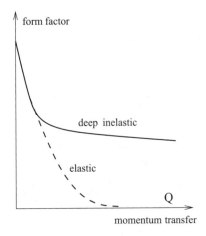

Figure 5.14. The form factor for deep inelastic electron–proton scattering. From the very slow decrease at large momentum transfer it was ultimately concluded that there are pointlike constituents within the proton.

One possible interpretation of these results is that the proton contains pointlike particles. This possibility had been proposed by J. D. Bjorken before the first experimental results were known, and well before they were publicly announced at the 1968 high-energy conference in Vienna.

From a purely theoretical analysis of current algebra results, Bjorken concluded that "we find these relations so perspicuous that, by an appeal to history, an interpretation in terms of elementary constituents is suggested." It should be remembered that in 1964, when Gell-Mann proposed quarks as the elementary—though not necessarily pointlike—constituents of hadrons, he was also motivated by results derived from current algebra.

Although Bjorken was a professor at Stanford, and had regular contact with experimentalists, his theoretical considerations did not impress his colleagues working on the experiment of deep inelastic scattering. Jerome Friedman, one of the principal collaborators in the experimental group, wrote the following about the results and the influence of Bjorken:

> These results were derived well before our inelastic measurements appeared. In hindsight, it is clear that they implied a point-like structure of the proton and the neutron and large cross sections at high q^2, but Bjorken's results made little impression on us at

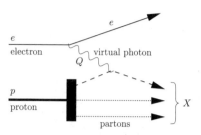

Figure 5.15. Schematic representation of deep inelastic scattering in the parton model. If the momentum transfer is large, the virtual photon resolves the proton into its point-like constituents (partons); the photon interacts with one of those partons (dashed line), whereas the others (dotted lines) fly on without interacting.

the time. Perhaps it was because his results were based on current algebra, which we found highly esoteric. Or perhaps we were very much steeped in the physics of the time, which suggested that hadrons were extended objects with diffuse substructures.

On another subject, that of *scaling behavior*, Bjorken was more successful in convincing his experimental colleagues of the importance of his theoretical results. Again, from analyzing current algebra results he found that—speaking in a simplified manner—deep inelastic scattering does not depend on both the variables Q (momentum transfer) and W (total energy of the hadronic state X), but only on the quotient of momentum transfer to energy, that is, Q/W.

The analysis of the data showed that scaling behavior really did hold. This was of special interest, since it could also be derived theoretically from a model proposed by Feynman in 1969: the parton model. In this model, hadron scattering is reduced to the interaction of pointlike constituents, called partons. Application of this idea to deep inelastic scattering is especially simple and is shown pictorially in Figure 5.15. The proton is a bunch of partons that fly along parallel paths and effectively do not interact with each other; the photon interacts with only one of them.

This model allows for a rather intuitive interpretation of the scaling behavior. First we introduce the *Bjorken scaling variable*, which is approximately given by $x_B \approx Q^2 c^2 / W^2$. In the parton model the scaling variable x_B is the fraction of the hadron momentum that is carried by the scattered parton (the dashed line in Figure 5.15). Therefore, in deep inelastic scattering one can measure the momentum distribution of the partons inside the proton. One should, however, not overrate this intuitive picture, for it is valid only for an observer who moves so fast with respect to the proton that the latter has, in the limit, infinite momentum. I will not further discuss this *infinite momentum frame*, which plays an important role in the interpretation of deep inelastic scattering. I have mentioned it only to show how far subnuclear phenomena are removed from our intuition, trained by everyday experience. Therefore, intuitive pictures normally explain some features of the phenomena very well, whereas they fail badly in other aspects.

It was of course tempting to identify the partons with quarks, but this was not the only possible interpretation. Feynman instead suggested that the partons corresponded to the *bare fields* of protons and pi mesons. Initially the parton model was by no means generally accepted; there were competing models that could also explain the data, although perhaps not quite as naturally. Furthermore, for the interpretation of partons as quarks there seemed to be an insurmountable obstacle: since no free quarks had been observed, they must be very tightly bound inside the hadrons, that is, they interact very strongly with each other. For the scaling behavior of the parton model, however, it is essential that the partons interact only very weakly. As has happened very often in the history of physics, this apparent contradiction was the germ for a new development in particle physics, as we will see in the next chapter.

It took a rather long time for the importance of the results of deep inelastic scattering to be fully recognized. It was not until 1990 that J. I. Friedman, H. W. Kendall, and R. E. Taylor were awarded the Nobel Prize, "for their pioneering investigations concerning deep inelastic scattering of electrons on protons and bound neutrons, which have been of essential importance for the development of the quark model in particle physics." Bjorken came away empty-handed.

For further investigations of structure functions, deep inelastic scattering experiments with neutrinos taking the place of electrons played an essential role. I already have mentioned that there were excellent neutrino beams in operation at the high-energy laboratories in Brookhaven, New York, and at CERN. With these beams, experiments similar to those at SLAC were performed, and the analysis of those experiments with the help of current algebra was decisive in the theoretical development described above. The analysis of electron scattering was safer, however, since in the 1960s there was not yet a satisfactory fundamental theory of weak interactions. Therefore, the principal results of neutrino scattering were in the field of weak interactions. I have already mentioned one important outcome, the existence of two distinct kinds of neutrinos. Another finding, the evidence for weak neutral currents, will be discussed in the next chapter, in Section 6.3.2.

6

The Standard Model of Particle Physics

Out of the approaches outlined in the last two chapters, a very consistent theoretical model, called the standard model of particle physics, was developed. It describes the strong, electromagnetic, and weak interactions as quantized gauge field theories. Only the oldest known interaction, that of gravity, is not incorporated in this model; to achieve this is still a great challenge.

6.1   Introduction

With the standard model, particle physics came nearer than ever before in the history of science to recognizing what keeps the inner world together. We now know the microscopic laws of strong, electromagnetic, and weak interactions. Many physicists today are disappointed not because the model works badly, but because it is so good that it leaves little room for new physics. The standard model is, like quantum mechanics and quantum field theory in general, the result of very close cooperation between experimental and theoretical physics. In this respect, two experimental discoveries were especially important. First, the results of deep inelastic scattering, which was discussed at the end of the previous chapter, and second, the discovery of weak neutral currents, which will be treated in some detail in this chapter.

As mentioned above, the standard model describes strong, electromagnetic, and weak interactions. The development of these branches was closely interwoven, and it turned out that all interactions are quantized gauge theories. The formal theory was more complicated for the weak interactions; in strong interactions, the physical situation was more enigmatic.

We saw in Section 1.4.2 that Fermi's theory of weak interactions was very successful in describing beta decay. In it, four fermions are coupled directly, as shown in Figure 1.13. The theory is unsatisfactory, however, since it does not allow one to calculate higher orders in perturbation theory without making additional assumptions: it is unrenormalizable. Furthermore, there is an old and, as it turned out, well-justified prejudice that interactions are not due to immediate contact but to the exchange of quantum fields. After the discovery of parity violation, interest in developing a consistent theory of weak interaction revived, and the idea of an intermediate boson that could convey the weak interaction was again pursued.

The electromagnetic interaction is conveyed by photons, the gauge bosons of U(1) symmetry. The idea that the weak and electromagnetic interactions are somewhat connected had been discussed in 1957 by Lee and Yang and by Schwinger. In 1958, S. Bludman proposed that the intermediate bosons of weak interactions are gauge bosons of an SU(2) symmetry. In that case they have to form a triplet consisting of a positively charged particle, a negatively charged one, and a neutral one. In such a theory, the observed decays, such as the beta decay of the neutron, are caused by an electrically charged virtual intermediate boson; see Figure 6.1(a).

The neutral boson leads to reactions such as those shown in Figure 6.1(b): a neutrino interacts with an electron from the shell of an atom, with the electron receiving a part of the (high) energy of the neutrino and flying off at a high velocity. The interaction shown by the graph of Figure 6.1(b) is mediated by the exchange of a neutral boson. One says that such processes are induced

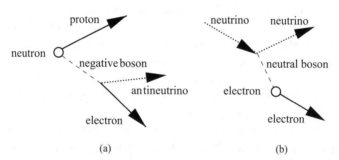

(a)   (b)

Figure 6.1. Charged and neutral weak currents. (a) Charged current: the neutron decays via a virtual *charged* intermediate boson. (b) Neutral current: a neutrino is able to knock an electron out of an atom via the exchange of a *neutral* intermediate boson.

by (weak) *neutral currents*. Note that in this terminology, the electromagnetic current is neutral, since it is due to the exchange of the (neutral) photon.

Bludman's model was extended in 1961 by Glashow, who proposed that the neutral intermediate boson of weak interactions mixes with the photon. For more on particle mixing, see Section 2.3. This idea was developed further by Weinberg and formulated in the framework of spontaneous symmetry breaking; we will discuss Weinberg's model in more detail in the next section.

There were other attempts to master the difficulties with higher orders of perturbation theory. In 1969, the prominent team of Gell-Mann, M. L. Goldberger, N. M. Kroll, and F. E. Low proposed a rather complex theory for the "amelioration of divergence difficulties in the theory of weak interactions," which removed at least some of the difficulties of Fermi's theory.

6.2   A Model for Leptons

In 1967, three years after the publication of the papers on the mechanism of spontaneous breaking of gauge symmetries (see Section 5.4), S. Weinberg proposed a model of leptons based on mass generation through spontaneous symmetry breaking. He assumed that the weak interaction is mediated by the three gauge bosons of an SU(2) symmetry. Electromagnetic interaction is conveyed by the gauge boson of the U(1) symmetry. Therefore, Weinberg posed the question: "What could be more natural than to unite these spin-one bosons into a multiplet of gauge fields?" Of course, he made use of the experimentally well-established fact that parity is maximally violated and therefore only left-handed particles participate in weak interactions. This was discussed in more detail in Section 2.8.

The neutrino is left-handed; the electron is massive, and therefore there is a left-handed as well as a right-handed electron. The left-handed electron and its neutrino, the *e*-neutrino, are grouped into a doublet of *weak isospin*. The neutrino is assigned the orientation $+\frac{1}{2}$, while the electron is given orientation $-\frac{1}{2}$. The right-handed electron forms a singlet and thus has weak isospin 0.

We provisionally consider only the gauge quanta of the two groups of the interactions, that is, U(1) and SU(2). The observed particles are mixtures of these. The gauge bosons of the symmetry group (SU(2)), designated by W^+, W^-, and W^0, couple only to the doublet of left-handed particles, whereas the gauge boson of U(1), called B, couples to both the right-handed and left-handed electrons. The W^0 and B bosons have the same quantum numbers and couple to the same particles. This can lead to particle mixing.

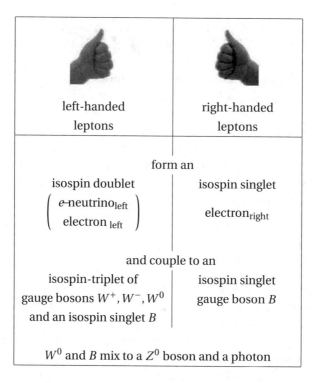

left-handed leptons	right-handed leptons

form an

isospin doublet	isospin singlet
$\begin{pmatrix} e\text{-neutrino}_{\text{left}} \\ \text{electron}_{\text{left}} \end{pmatrix}$	electron$_{\text{right}}$

and couple to an

isospin-triplet of gauge bosons W^+, W^-, W^0 and an isospin singlet B	isospin singlet gauge boson B

W^0 and B mix to a Z^0 boson and a photon

Figure 6.2. Summary of Weinberg's model of leptons.

The degree of mixing is expressed by an angle called the *electroweak mixing angle* (or *Weinberg angle*). The result of the mixing of B and W^0 are the photon and the neutral intermediate boson Z^0. Figure 6.2 summarizes the results schematically.

For the muon and its neutrino, the procedure is the same: one forms a doublet of the left-handed negative muon and the mu neutrino and a singlet of the right-handed muon. In Figure 6.2 we have only to replace the electron by the muon. We will return soon to the underlying ideas when we discuss the standard model in its full glory in Section 6.10.

One can extend the Gell-Mann–Nishijima relation between charge, isospin, and hypercharge (see Section 2.2) to leptons by assigning the (weak) hypercharge $Y = -1$ to the left-handed particles and $Y = -2$ to the right-handed ones. The relation is then the same as for flavor SU(3):

$$Q = I_3 + \tfrac{1}{2}Y.$$

Here again, Q is the electric charge and I_3 the (weak) isospin orientation. For antiparticles, the hypercharge and I_3 have opposite signs.

In the first years after its publication, Weinberg's paper aroused little interest. Indeed, in the first three years after its appearance it was not even cited by its author. The first registered citation came in 1974. This seems strange for a paper that later became so famous (by 2007 it had been cited more than 6,000 times) and was written by such a prominent researcher. But its early fate is understandable, since only two predictions were made in the paper.

One prediction was the existence of neutral currents, but there was no experimental evidence for them at the time, and furthermore, the prediction was not new: Bludman had made it already in 1958. The other prediction, however, was truly original, since it was based on mass generation through spontaneous symmetry breaking, the so-called Higgs–Kibble dinner (see Section 5.4). With this theory, Weinberg was able to predict the masses of the gauge bosons, whose precise values can be determined only if the above-mentioned mixing angle is known precisely. But even without knowledge of the angle, Weinberg could give lower bounds for the masses. The W^+ and W^- bosons should have a mass of at least 40 GeV/c^2, and the neutral one, the Z^0, a mass of at least 80 GeV/c^2. This prediction was rather daring, since the gauge bosons had to be at least ten times as massive as any known elementary particles (including all resonances), more massive even than the atoms of medium-heavy elements such as calcium. I still remember the laughter from members of the MIT theory division when, in the lunchroom, someone reported on Weinberg's paper with those mass values.

The theoretical foundation was based on hope rather than on solid theorems. Weinberg admitted this openly in his paper:

> Is this model renormalizable? We usually do not expect non-Abelian gauge theories to be renormalizable if the vector–meson mass is not zero, but our Z_μ and W_μ get their mass from the spontaneous breaking of the symmetry, not from a mass put in at the beginning. Indeed, the model Lagrangian we start from is probably renormalizable, so the question is whether this renormalizability is lost ... by our redefinitions of the fields.

Likewise, A. Salam, who in a paper with ideas similar to those of Weinberg had a few more words about renormalization, could only offer hope but not give proofs. Finally, it should be remarked that the intended unification of electromagnetic and weak interactions was not achieved. In the case of a true unification, the electroweak mixing angle would be determined by theory.

Another weak point of the theory was that it could be applied only to decays of leptons, that is, muons. Decays of hadrons were not treated by Weinberg. It was an obvious step to apply the theory to quarks as well, and Salam expressed hope in this direction. It should be noted, however, that in 1967, quarks were far from being integrated into quantum field theory.

An important point was the existence of processes mediated by exchange of the neutral intermediate boson, that is, the existence of neutral currents. There were no indications that they existed, though there were strong indications that they did not exist at all, or at least were strongly suppressed, in decays in which strangeness changed.

If one assumes that quarks couple like leptons to intermediate bosons, there was no reason that the neutral boson Z^0 should not couple to the neutral K meson. Since the Z^0 couples to an electron–positron pair, one should expect the decay of a neutral K meson into, for example, a neutral pi meson and an electron–positron pair. However, such a decay has never been observed in spite of very thorough investigations of neutral K decays. From this, it follows that the coupling of neutral currents to quarks was more complicated than the coupling to leptons—if neutral currents existed at all.

On the way to the standard model as we know it today there were some further obstacles. For the purely leptonic decays it was the missing proof for renormalizability. This proof was completed by 't Hooft and Veltman in 1971; see Section 5.3. This was an important step forward, but evidence for neutral currents in purely leptonic processes, as shown in Figure 6.1(b), was missing. It was clear that the detection of decays with neutral currents would be difficult.

For hadronic processes the prospects were even bleaker. First, there was no respectable field theory of strong interactions, but even if one closed one's eyes to that, there was the problem with strangeness-changing neutral currents. One had to explain why the decay of a neutral K meson into a pi meson, an electron, and a positron was not observed. We will devote our attention to this problem before we come to the experimental search for neutral currents.

6.3   Weak Currents

6.3.1   A Miracle Vanishes

Presumably, the results of deep inelastic scattering made the quark model more attractive for physicists interested in quantum field theory. In 1970, S. Glashow,

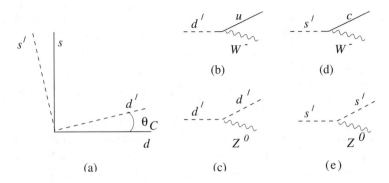

Figure 6.3. (a) The construction of the mixed states s' and d' from the s and the d quarks. The mixed states couple directly to the intermediate bosons W^- and Z^0, as shown in graphs (b)–(e).

J. Iliopoulos, and L. Maiani addressed the problem of strangeness-changing neutral currents. This problem must have seemed rather academic to many phenomenologically oriented physicists: it existed only if there were neutral currents, for which there was no experimental indication. For rigorous theoreticians, on the other hand, this investigation was suspicious for two reasons: one had no consistent theory for the quark model, nor had the renormalizability of the weak interaction theory with heavy intermediate bosons been proven. But it turned out that the investigations of Glashow, Iliopoulos, and Maiani paved the way to the standard model.

We start with a quark model without committing ourselves to details. Glashow, Iliopoulos, and Maiani, commonly abbreviated GIM, did not even stipulate whether the quarks have fractional charge, as in the model of Gell-Mann and Zweig, or integer charge, as in the model of Han and Nambu; see Sections 4.3 and 4.4. In this brief discussion I confine myself to fractionally charged quarks, and I will use the modern names u, d, and s quark (GIM used $\mathscr{P}$, $\mathscr{N}$, and λ).

To describe developments in weak interactions I have to go further back. In 1963, shortly after the introduction of flavor symmetry in order to classify hadrons, N. Cabibbo proposed a classification of weak interactions according to this symmetry. This theory is very successful and is easily incorporated into the quark model. One has only to assume that for weak interactions, not the u and d quarks individually, but only some superposition (mixture) of them, is relevant.

We can explain the mixing in the following way: we represent the d and s quarks in a plane by lines at right angles; see the solid lines in Figure 6.3. In weak interactions it is not the d and s quarks that are directly relevant, but their superpositions d' and s', represented by dashed lines. The rotation angle θ_C is the so-called Cabibbo angle. This angle was determined from decays of strange particles to be about $13°$, that is, rather small. The d' mixture, relevant for weak interactions, has thus a large component of d and a small component of s. In the quark model of weak interactions, the left-handed component of the d' state plays the same role for hadronic decays as the electron for leptonic decays, whereas the role of the neutrino is taken by the u quark. The weak isospin doublet of quarks that couples to the intermediate bosons W and Z is therefore not $\binom{u}{d}$ but $\binom{u}{d'}$. The resulting interaction terms are shown in Figures 6.3(b) and (c).

The strangeness-changing decay of a neutral K meson into a positive pi meson, an electron, and an antineutrino is displayed in Figure 6.4(a). It is a consequence of the interaction shown in Figure 6.3(b), since the s quark, a constituent of the K meson, is contained in the mixture d'. The coupling of the neutral intermediate boson Z^0 to the d', Figure 6.3(c), leads to a transition of the s to the d, since both the s and d quarks are contained in the mixture d'.

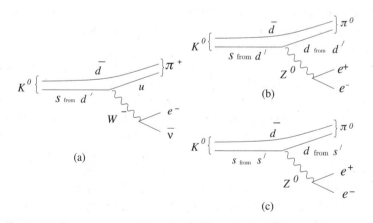

Figure 6.4. Decays of K mesons in the quark picture. The neutral K meson (K^0) consists of an s quark and an anti-d quark ($\bar{d}$). The positively charged pi meson (π^+) consists of a u quark and an anti-d quark ($\bar{d}$), and the neutral pi meson (π^0) contains a d quark and an anti-d quark ($\bar{d}$). The contributions of (b) and (c) have opposite signs and cancel each other.

Therefore, one expects a decay of the neutral K meson into a neutral pi meson and an electron–positron pair. This should happen with similar probability to that of the decay of a charged K meson into a positively charged pi meson, an electron, and an antineutrino. Experimentally, however, the decay of a neutral K meson into a neutral pi meson and an electron–positron pair has never been observed; it must be at least 1,000 times less probable than the decay into a charged pi meson.

Glashow, Iliopoulos, and Maiani found an elegant solution by pulling out a new quark, like a rabbit out of a hat. For the leptons we have two *families* that couple to the intermediate bosons, namely the electron with its neutrino and a corresponding muon family. Under the rotation of the d quark into the d' quark, the s quark becomes a mixture s', which in Figure 6.3 points weakly upward to the left. This mixture consists mainly of the s quark, but it has also a small admixture of the d quark; this admixture is negative, since s' points to the left. This mixture s' had no natural place in the old Cabibbo theory. Glashow and his collaborators assumed in addition that this s' mixture participates in the weak interaction; its left-handed part should be the lower component of a weak isomultiplet, in analogy to the d' in the known multiplet.

Since an upper component of the new multiplet was unknown, they postulated the existence of a fourth quark, having the same charge as the u quark but that must have some property that distinguishes it from the u quark. The three authors fittingly called this property of the magically produced quark "charm."

The property of charm is very similar to strangeness. Like the latter, it does not change in strong interactions, but a quark with charm, called a c quark, can decay weakly through the coupling $c - s' - W^-$ (Figure 6.3(d)) into an s quark or a d quark. Furthermore, there is the coupling to the neutral Z^0 boson; see Figure 6.3(e). The different signs of the mixtures of the s quark into the d' and of the d quark into the s' result in a cancellation of the two terms in Figure 6.4(b) and (c). This is why the decay does not take place or is at least strongly suppressed. The four doublets of fermions participating in weak interactions are collected in Table 6.1.

If these considerations had not been later confirmed by experiment, one could have easily made fun of them. In order to explain why an unobserved intermediate boson does not give rise to an unobserved decay, one introduces a further unobserved particle, the c quark. And to top it all off, this was all assembled within a somewhat shaky theoretical framework: the quark model.

This theory of the c quark, named the *GIM mechanism* after the initials of its authors, solved all the problems with neutral strangeness-changing cur-

$$\begin{pmatrix} e - \text{neutrino}_{\text{left}} \\ \text{electron}_{\text{left}} \end{pmatrix} \qquad \begin{pmatrix} u - \text{quark}_{\text{left}} \\ d' - \text{mixture}_{\text{left}} \end{pmatrix}$$

$$\begin{pmatrix} \mu - \text{neutrino}_{\text{left}} \\ \text{muon}_{\text{left}} \end{pmatrix} \qquad \begin{pmatrix} c - \text{quark}_{\text{left}} \\ s' - \text{mixture}_{\text{left}} \end{pmatrix}$$

$$d' = \cos\theta \cdot d + \sin\theta \cdot s$$
$$s' = -\sin\theta \cdot d + \cos\theta \cdot s$$
$$\sin\theta \approx 0.22, \ \cos\theta \approx 0.95$$

Table 6.1. The four doublets participating in weak interactions. The d' and s' mixtures are superpositions of d and s quarks.

rents, not only the decay just considered. In particular, one could show that the fourth quark was also necessary in quantum corrections. From this, one could estimate the mass of the c quark to be at most about $2\,\text{GeV}/c^2$.

In 1970, when this theory was being assembled, there were many indications that quarks do not exist as free particles. Therefore, one did not expect to observe the c quark directly, but instead, hadrons containing c quarks. The lightest meson containing a c quark can decay only weakly, since charm is conserved in strong interactions; Glashow and his collaborators predicted for it a mass of about $2\,\text{GeV}/c^2$ or less and a lifetime of about 10^{-13} seconds. From these values one could draw conclusions on masses and lifetimes of hadrons containing a c quark. Thus there were solid predictions growing out of these speculations, and they made the GIM mechanism very attractive, at least for theoreticians. In 1975, a review article appeared in which all the essential properties were predicted, especially the existence of a bound state of a c quark and its antiparticle.

The c quark had another beneficial effect. The proof of renormalizability of a spontaneously broken gauge theory still had a snag: for the proof, the symmetry of the fundamental equations is essential. However, this symmetry is broken by quantum corrections, the so-called anomalies (see Section 5.5). The left-handed fermions in the electroweak theory give rise to anomalies. They lead to an explicit breaking of the gauge symmetry and therefore jeopardize renormalizability.

But here the charm of the fourth quark offered a solution. The contributions of the quarks and those of the leptons to the anomaly have different signs.

Because of that, there is no anomaly in a complete family: the anomaly of the electron and that of its neutrino are canceled by those of the u and d quarks. The muon anomaly, on the other hand, is only partially canceled by that of the strange quark alone. But after the introduction of the c quark it is canceled exactly. The same game was played again later; we will come to this in Section 6.8.

A group of fermions in which the anomalies cancel one another is called a *family*. It contains a charged lepton and its neutrino and two quarks with electric charges $+\frac{2}{3}$ and $-\frac{1}{3}$, respectively. Up to now, we have heard of two families. The first consists of the electron and its neutrino and the u and d quarks, while the second consists of the muon and its neutrino and the s and—at the time, around 1970, still hypothetical—c quarks. The quarks of different families mix under the weak interaction.

6.3.2 The Needle in the Haystack Is Found

Independent of all theoretical investigations, it was, of course, a critical task to look for events due to neutral currents. Fortunately, there were just the right prerequisites at CERN to detect such events. There was a very good beam of mu neutrinos from the decay of positive mesons and a beam of anti-mu neutrinos from the decay of negative mesons. Furthermore, there was the bubble chamber "Gargamelle," which was just the right detecting device for neutrino reactions. As the name indicates—Gargamelle was the gigantic mother of the giant Gargantua from Rabelais' novel *Gargantua and Pantagruel*—the bubble chamber was enormous: it was 4 meters long with a diameter of 1.9 meters and weighed in at 20 tons. Such a large size was necessary if there was to be any chance at all of detecting neutrino reactions.

Recall that Pauli was afraid that "the foolish child of his life crisis" would never be detected, and that Reines first thought he would need an atomic bomb explosion in order to prove the existence of neutrinos directly.

Since the chance of detecting a reaction increases proportionally with the volume, it is no wonder that one needs a detector of gargantuan proportions. The chamber was filled with Freon, a compound of carbon, fluorine, and chlorine, which under a pressure of 20 atmospheres is liquid at room temperature; therefore, it is much less volatile and dangerous than a hydrogen-filled chamber. For the analysis of purely hadronic events it was less well suited than a hydrogen chamber, since reactions with complex nuclei cannot be analyzed so unequivocally.

But the gigantic chamber was most suitable indeed for detecting neutrino reactions. Besides the sheer size, the high density of Freon helps, since there are many more nucleons and electrons per unit volume in liquid Freon than in liquid hydrogen. The binding of the nucleons in nuclei does not interfere with the analysis of high-energy neutrino reactions, since the range of weak interactions is very short.

In a priority list for neutrino experiments at CERN, drawn up in 1968, the search for neutral currents was only eighth out of ten. At the time, the tests of current algebra seemed more urgent. Nevertheless, there were enough undaunted physicists in the Gargamelle Neutrino Collaboration to become involved in this difficult experiment. They later received additional motivation when in 1971 't Hooft and Veltman finished the proof of the renormalizability of nonabelian gauge theories.

To investigate leptonic neutral currents means to look for traces in the bubble chamber that appear out of nowhere, as indicated schematically in Figure 6.1(b). A neutrino, invisible in the chamber, hits an electron, also invisible, since it is inside the shell of an atom. If the neutrino has sufficiently high energy, the electron is knocked out of the atom and leaves behind a trace, while the scattered neutrino flies off invisibly.

Somewhat simpler to see are neutral current interactions with nucleons in atomic nuclei. Here several visible traces develop, seemingly from nothing. They originate at the point where the neutrino has hit the nucleus. Such events, however, are easily confused with the abundant nuclear reactions of neutrons. Charged currents are much easier to detect. In this case, at each reaction a muon is created and easily identified.

At the end of 1972, after approximately 100,000 pictures had been scanned, one event was found in which an anti-mu neutrino had hit an electron that left a clearly visible trace. Fortunately, the probability of confusing this reaction with other reactions is extremely small. Therefore, this single event proved the existence of neutral currents. In the remaining 1.3 million pictures, two further events of this type were found, a number that agreed with that expected from the model of leptons. From these three events one could estimate the electroweak mixing angle and the masses of the W and Z^0 bosons. For the W boson one obtained a mass range between 50 and 120 GeV/c^2, while for the Z boson, the range was between 77 and 126 GeV/c^2.

By a very careful analysis, it was possible to separate the hadronic neutral-current events from those induced by neutron reactions, and in 300,000 pictures, 166 events were detected. This allowed a much better determination of the mixing angle and thus a better prediction for the boson masses. For the charged gauge bosons $W^\pm$, the mass turned out to be in the interval 60 to

70 GeV/c^2, while for the neutral Z^0 the range was 77 to 83 GeV/c^2. Both groups published these results in 1973 in the same issue of *Physics Letters B*.

To come to the point, I give the present-day results: 80.40 GeV/c^2 for the charged $W^\pm$ bosons and 91.188 GeV/c^2 for the neutral Z boson. Getting slightly ahead of our story, I note that the 1979 Nobel Prize in physics was awarded to Glashow, Salam, and Weinberg, though at this time the intermediate bosons had not yet been established experimentally. For the proof of renormalizability of gauge theories, 't Hooft and Veltman received the Nobel Prize only in 1999.

6.4   The Strong Interaction Becomes Dynamic

In Section 4.4 we learned that in 1965, Han and Nambu proposed that quarks have an additional quantum number, which they called color, thereby achieving two things: first, they could overcome the difficulties with the exclusion principle, and second, they could assign the quarks an integer charge. The three colors of the quarks should transform according to the SU(3) symmetry group; in the following we shall call this symmetry color SU(3) in order to distinguish it from the flavor SU(3) (the eightfold way).

Gell-Mann wrote later that he was never impressed by the Han–Nambu quarks and their integer charges, since he had never believed that quarks could be isolated as free particles. He should have been more impressed by a dynamical model that was indicated in the paper of Han and Nambu by a single sentence. In this model, the force between the quarks is mediated by a vector meson that transforms according to an octet representation of color SU(3). This might have brought him in a direction that he pursued only later with H. Fritzsch. In 1971, the two took up the idea of colored quarks. However, they considered fractionally charged (Gell-Mann–Zweig) quarks and assumed that isolated hadronic states must always be color neutral, that is, transform as color singlets under color SU(3). With that, they postulated that the color symmetry is an unbroken symmetry and that all physically realized states are invariant under this transformation. One also says that hadrons are *color neutral* or *colorless*.

An important argument supporting the assumption that the quarks come in triples came from the decay of the neutral pi meson. A simple graph for that decay is the triangle graph of Figure 6.5. It describes the creation of a virtual quark–antiquark pair that in turn annihilates into two photons. By weighty theoretical arguments, related to the Goldstone theorem (Section 5.3) as well as to

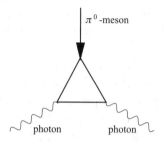

Figure 6.5. The Feynman graph for the decay of a neutral pi meson (π^0). The inner lines represent virtual quarks. There is one such graph for each flavor and for each color of the internal quark.

anomalies (Section 5.5), this simple graph should give a good value for the decay probability of the pi meson. If one calculates the graph with the usual fractionally charged quarks, the resulting decay probability is much smaller than what is observed. But if each quark occurs in three colors, there are three times as many graphs, and the calculated value then agrees very well with the experimental one.

In 1972, Gell-Mann presented the model in conferences, and in 1973, a paper by H. Fritzsch, M. Gell-Mann, and H. Leutwyler appeared with the title "Advantages of the Color Octet Gluon Picture." In this paper, the authors propose that the color SU(3) according to which quark colors transform is a *local* gauge theory, that is, a Yang–Mills theory. From this it necessarily follows that there are eight gauge bosons interacting with the quarks.

These gauge bosons, called *gluons*, cause the strong interaction in the same way as the gauge bosons of quantum electrodynamics (QED), the photons, cause the electromagnetic interaction. Thus the theory of strong interactions was of a similar structure to that of QED: an unbroken local gauge theory. But in contrast to the latter, the gauge bosons carry color charge.

In analogy to QED, in 1974, Gell-Mann coined the name *quantum chromodynamics*, abbreviated QCD, for this field theory of strong interactions. Since the gauge bosons of QCD, the gluons, are members of an octet, they are not invariant under color SU(3) transformations and should therefore not be directly observable, just like quarks. Gell-Mann and coworkers saw it as positive that the quarks and gluons were unobservable in their model. In this way, there was no contradiction with the principle of nuclear democracy, which postulates that all hadrons have to be treated on the same footing.

As a field theory, QCD was as respectable as QED (even more respectable, as it later turned out), since renormalizability had been shown already in 1971 by 't Hooft and Veltman. Strangely, this is not mentioned in the paper of Fritzsch, Gell-Mann, and Leutwyler.

The model was elegant and theoretically respectable, but it had not been used to perform quantitative calculations. The three authors even took into account the possibility that local gauge symmetry does not determine the dynamics completely and that it had to be modified at short distances in order to explain Bjorken scaling (Section 5.8) and at long distances in order to explain *confinement*. The term "confinement" was given to the phenomenon that quarks and gluons are apparently permanently bound (confined) in hadrons and can never be isolated. This somewhat unsatisfactory situation changed only with the discovery that QCD has a special property, called asymptotic freedom. This will be discussed in the next section.

6.5 Running Coupling and Asymptotic Freedom

In QED, something called *vacuum polarization* played an important role. It is the process by which a photon can transmute into a virtual electron–positron pair, that is, the photon polarizes the vacuum into two different charges. The simplest contribution to vacuum polarization is shown as a Feynman graph in Figure 6.6(a).

Soon after the first discussion of vacuum polarization by Dirac and Heisenberg in 1934, R. Serber and E. A. Uehling discovered that the interaction between two electric charges is modified by it. The contribution of vacuum po-

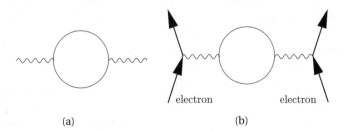

(a)   (b)

Figure 6.6. (a) A contribution to vacuum polarization in QED: a (virtual) photon dissociates into a virtual electron–positron pair. (b) Contribution of vacuum polarization to electron scattering; it leads to an increase of the renormalized charge at small distances.

larization to scattering of an electron on an electron is shown as a Feynman graph in Figure 6.6(b). It gives rise to the following phenomenon: for smaller distances, the interaction increases faster than would have been expected from classical electrodynamics. This implies that for reactions that take place at very short distances, the charge is larger than that for reactions in which a larger scale is relevant. Qualitatively this can even be understood by considering that a classical polarization leads to the formation of small dipoles in a medium, in this case the vacuum. The positive poles of the dipoles (the virtual positrons in our case) are attracted by the electrons, and therefore the dipoles screen the charge; it decreases with increasing distance faster than without screening.

Inquiry into these considerations was resumed in 1953 in the framework of quantized field theory by Gell-Mann and F. E. Low, as well as by E. C. G. Stückelberg and A. Petermann; the result was one of the most important tools in quantum field theory, the *renormalization group*. It is impossible to give here more than a rather superficial explanation of this concept, but I will nevertheless mention two seemingly quite different approaches.

One approach is directly related to the method of renormalization, that is, the taming of infinities. I already have mentioned that in renormalized perturbation theory, the "bare" parameters, such as charge and mass, are first left open, and the renormalized quantities are defined by comparing with experiment at a certain scale, the so-called renormalization scale. It is easy to see that the renormalized parameters depend on the scale. But on the other hand, directly measurable quantities such as cross sections had better not depend on this scale. This requirement makes it possible to determine the dependence of the renormalized quantities on the scale by what is called the renormalization-group equation. In 1970, this development was given a certain degree of conclusiveness by C. G. Callan and K. Symanzik in the equations named after them. From these equations one can see directly how the contribution of the graph of Figure 6.6(a) leads to an increase in the renormalized charge at smaller distances.

A less formal approach is due to L. P. Kadanoff, whereby the renormalized constants have a rather intuitive meaning. The renormalized constants account for all the effects important at distances smaller than the renormalization scale. Only for renormalizable theories is it possible to take into account all small-distance effects in a simple renormalized constant. In both approaches, the renormalized charge increases with diminishing distance in QED.

Let us return to high-energy physics. In the parton model it was assumed that at small distances, the pointlike constituents of hadrons, the partons, practically do not interact. If one wanted to identify the quarks with the partons, one had to assume that the interaction was very weak at small distances, but strong at large distances. Only in this way could one explain scaling as well as the fact that quarks had not been observed as free particles.

Therefore, in QCD, if it is the theory of strong interactions, the behavior should be opposite that of QED, where the coupling is small at large distances but increases with smaller ones. This seemed impossible, since the intuitive interpretation of the vacuum polarization predicts an increase at smaller distances, and more rigorous arguments based on unitarity seemed to go in the same direction. One might therefore think that the discovery that QCD has just this behavior, precisely opposite to that of QED, fell like a bombshell. Yet this was not the case, and I will tell the story of this discovery before I continue with the physical consequences.

In 1965, two Russian physicists, V. S. Vanyashin and M. V. Terentev, calculated the vacuum polarization for a theory with massive charged particles with spin 1, effectively a directly broken SU(2) gauge theory. This theory is not renormalizable, but the lowest quantum correction can be calculated. The two authors found that the behavior of vacuum polarization was opposite to that in QED; that is, the renormalized coupling decreases for smaller distances. Another name for this effect is *antiscreening*. The authors found that this effect "seems extremely undesirable." Though their paper was published in a respected journal and was translated into English, nobody, including the authors, thought to apply antiscreening to the parton model.

The second act of the drama is even stranger. After he had proved renormalizability, the young, brilliant physicist 't Hooft had also calculated the renormalized color charge, that is, the coupling in QCD. He had found that it decreases with smaller distances. In contrast to the discovery mentioned above, this was now an effect in a fully consistent theory and already interesting in itself.

At a small conference in Marseille in June 1972, 't Hooft met the physicist K. Symanzik, from Hamburg. The latter was a great specialist in these questions, and he told 't Hooft that in some unrealistic theories he had found such antiscreening, and he mentioned that this could be relevant for the parton model. Now 't Hooft reported his results from QCD, and Symanzik was surprised and presumably also skeptical. He advised his young colleague, "If this is true, it will be very important, and you should publish this result quickly, and if you won't, somebody else will." In his seminar in Marseille, 't Hooft mentioned his result, but he did not follow Symanzik's advice to publish it as fast

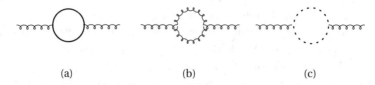

(a)   (b)   (c)

Figure 6.7. Contributions to vacuum polarization in QCD. The solid lines represent quarks (fermions); the coiled lines, gluons; and the dashed lines, ghosts (see Figure 5.4). Graph (a) is analogous to the vacuum polarization in QED (Figure 6.6). The contributions (b) and (c), typical for QCD, lead to antiscreening, that is, decrease of the color charge for smaller distances.

as possible. So this peculiar behavior had to be discovered yet a third time—in 1973 by D. J. Gross and F. Wilczek and independently by H. D. Politzer. They realized the importance for the parton model, and in 2004 were awarded the Nobel Prize for "the discovery of asymptotic freedom in the theory of the strong interaction."

In QCD, the strength of the charge decreases with smaller distances and approaches zero in the limit of zero distance (or equivalently, infinite energy). In this limit, the theory seems to behave like a free theory, and therefore the name "asymptotic freedom" was coined for this antiscreening effect. In retrospect, one may wonder why physicists were apparently so slow to accept this antiscreening effect. The different behavior from that of QED can be understood easily. In QCD, the gauge bosons, the gluons, carry color charge and therefore couple to each other, in contrast to photons. Figure 5.4 shows the graphs for the purely gluonic couplings. This has important consequences for the vacuum polarization: the gluon can also dissociate into two gluons, not only into a fermion–antifermion pair, like the photon. Two gluons with the same color charge can attract each other, and this causes antiscreening: the renormalized color charge decreases with decreasing distance.

The Feynman graphs for the contributions of gauge bosons to vacuum polarization in QCD are displayed in Figure 6.7. The coiled lines represent gluons; the straight lines, quarks. The graph in Figure 6.7(a) occurs also in QED, but not the graphs in Figures 6.7(b) and (c), which are typical for a nonabelian gauge theory such as QCD. In the graph in Figure 6.7 (b) the gluon dissociates into two virtual gluons; the third diagram, Figure 6.7(c), shows the dissociation into a ghost and an antighost. The quantum fields for ghosts do not occur in the classical theory, but have to be introduced in order to obtain a consistent perturbation theory, especially to guarantee conservation of probability. Their

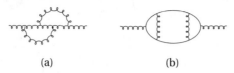

(a)   (b)

Figure 6.8. Higher-order contributions to vacuum polarization in QCD.

appearance is one of the reasons for the long time delay between the establishment of classical nonabelian gauge theory by Yang and Mills (1954) and the final proof of renormalizability by 't Hooft and Veltman (1971).

After one has calculated the expressions corresponding to the graphs in Figure 6.6, one can use the renormalization group techniques in order to determine how the interaction changes with distance. For more precise knowledge of this dependence, one has to calculate more graphs as well as more complicated ones. Some examples are shown in Figure 6.8.

In order to give expressions like "the weaker the interaction" some quantitative meaning, we introduce the strong coupling constant g_s (the subscript s stands for strong). It is the color charge, and it determines the interaction in QCD in the same way that the electric charge determines the interaction in QED. In quantum corrections, the quantity $\alpha_s = g_s^2/(4\pi)$ always appears.

In the interaction corresponding to the graphs in Figure 6.7, two elementary interactions (vertices) occur, and therefore the expressions corresponding to these graphs are proportional to g_s^2 and thus to α_s. Besides those terms, there are more complex ones that are quadratic or of a higher power in α_s. The contribution from Figure 6.8(a) contains four vertices and is therefore proportional to α_s^2; graph (b) is proportional to α_s^3.

The smaller the coupling, the less important are the higher-order contributions. Let us assume that at a certain distance, $\alpha_s = 0.1$. Then all contributions of the perturbation series not displayed in Figure 6.7 are multiplied by at least an additional factor $\alpha_s = 0.1$. We therefore expect these expressions to amount to only about 10% of the lowest-order contributions that are shown in Figure 6.7

Figure 6.9 shows the coupling $\alpha_s = g_s^2/(4\pi)$ as a function of distance. The solid curve is calculated to lowest order; that is, only the graphs shown in Figure 6.7 have been taken into account. The curve with long dashes takes into account the next order, that is, it also considers the graphs that are proportional to the square of α_s. An example of such a diagram is shown in Figure 6.8(a). The curve with short dashes includes the cubic terms, obtained in truly heroic

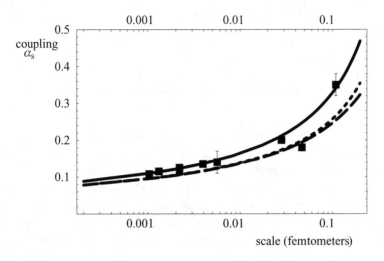

Figure 6.9. The running coupling in QCD. The solid curve is the prediction from first (lowest) order, the long-dashed curve includes the second order, and the short-dashed curve includes the third order in perturbation theory. The data points are taken from experiments that receive their main contribution at the distance indicated on the x-axis.

calculations. That a further order, terms in α_s^4, will ever be calculated is doubtful unless someone comes up with a completely new idea as to how to calculate the nearly astronomical number of complicated graphs.

One sees from Figure 6.9 that up to a distance of about 0.2 fm, corresponding to an energy of more than 1 GeV, the lowest-order calculation agrees qualitatively quite well with the second-order and third-order calculations, and the second and the third are already nearly equal. This gave theoreticians working in the field of strong interactions a new feeling of confidence in performing quantitative calculations in strong interactions. We will see more examples of this in the next three sections.

Figure 6.9 shows data points as well. These are values for the coupling α_s extracted from experiments in which a certain distance is particularly important. I will not go into detail or discuss how the theoretical curves should be compared with the experimental points. I simply want to stress that the "running of the coupling" has also been observed experimentally.

The curves show clearly the increase of the coupling with increasing distance, but the larger the coupling, the less reliable the calculations are, since higher orders in the coupling become more important. One can see in the figure that the difference between the different orders grows. It is not excluded that at great distances the coupling becomes so large that it is impossible to knock the quarks out of the hadrons. This effect is called *infrared slavery*, in distinction to asymptotic freedom. But in contrast to asymptotic freedom, infrared slavery is rather more wishful thinking than a theoretically well-founded concept. Fortunately, there are other approaches to investigating QCD at large distances, as we will see in Section 6.7.

The strength of the coupling can be expressed by a scale that is typical for QCD, called lambda QCD (Λ_{QCD}). It has the value of about 1 femtometer, or in energy units 200 MeV. At distances smaller or energies larger than this scale, the coupling is small, and perturbation theory can be applied meaningfully. At distances that are comparable or longer, perturbation theory breaks down. One can make the following analogy: with a microscope of a resolution power better than 1 femtometer, one can see quarks and gluons inside the hadrons, while with a microscope having lower magnification, one sees only hadrons.

Not only does asymptotic freedom explain the peculiarities of deep inelastic scattering, it also brings an additional fundamental advantage. I mentioned in Section 3.1 that Landau and Pomeranchuk had put forward serious arguments against QED and quantum field theory in general. They concluded that in a consistent QED theory the charge had to be zero, that is, there should be no electromagnetic interaction at all. The basic idea of their argument was this: renormalization at a certain distance means that all effects at smaller distances are taken into account in the renormalized quantities. If the charge increases with decreasing distance, this implies that the effects at shorter and shorter distances become more and more important, and since we can never reach distance zero, we will never be able to explore the full content of the theory, unless the coupling is zero to begin with.

This argument, which was supposed to be general, does not, however, hold for asymptotically free theories. Here the coupling decreases with shorter distances, and therefore the short-distance effects become less and less important. Asymptotically, we can very well explore the full theory, at least at short distances.

It has rightly been pointed out that Landau's arguments against QED are based solely on perturbation theory and that nonperturbative effects might alter the short-distance behavior. But nonperturbative numerical calculations have strongly supported Landau's conjecture. The use of QED as an effective theory at energies presently reachable as well as in the distant future is untouched by these fundamental considerations. We will return to nonasymptotically free theories at the beginning of Section 7.3.

In the next section we will consider some examples of the application of perturbation theory to short-distance phenomena.

6.6   Quantitative Calculations in Strong Interactions

We first focus our attention once again on deep inelastic scattering. We have seen in Section 5.8 that the parton model gave an explanation of the most salient features of deep inelastic scattering, that is, scattering of electrons on protons with high momentum transfer and high inelasticity. The paradoxical features of this model mentioned above were explained by the property of asymptotic freedom in QCD. But QCD as a gauge theory is much more powerful than the simple parton model. In the parton model, the interaction between the constituents is zero, while in QCD it only decreases with decreasing distance; it is never zero. Therefore one expects deviations from the predictions of the parton model; the reaction should depend not only on the Bjorken variable x_B, the ratio of momentum transfer to energy, but also directly on the momentum transfer.

This dependence can be calculated in QCD theoretically. One cannot calculate the absolute values of the cross sections, but only their dependence on the momentum transfer Q. In order to do so, one has to separate the short-range effects, which can be calculated with perturbation theory, from the long-range effects, which cannot. For that purpose, K. Wilson developed the *operator product expansion*. With it and the renormalization group technique one can precisely calculate the cross section (or structure functions) for any large value of the momentum transfer, if it is known for one large fixed value. The drastic deviations from naive scaling that were later observed will be discussed in connection with Figure 6.12.

In the early 1970s, however, the experiments were not so precise, and the data did not cover so wide a range that the results could have been universally convincing. The conviction that QCD is *the* theory of strong interactions gained acceptance only slowly.

Another application of QCD was the annihilation of highly energetic electrons and positrons. In storage rings, one can shoot a beam of electrons against a beam of positrons and detect the reaction products. Theoretically, this can be described by assuming that the electron and positron annihilate into a virtual photon and that this photon dissociates into all possible states. The processes in which a lepton pair—electron and positron or positive and negative muons—again appears can be calculated very precisely in QED.

But the virtual photon can also dissociate into hadronic states, and here the parton model and QCD also allow quite precise predictions. The annihilation process happens in a very small volume, and therefore the quark–parton model can be applied. One therefore assumes that for hadronic states a quark–antiquark pair is created. However, one does not observe the quarks, only the hadrons, and therefore there must be a further reaction in which the virtual quarks transform into hadrons. One calls this last stage of the process *hadronization*. This process is not understood even today, but there are good reasons to believe that certain quantities are at most only weakly affected by this hadronization process.

In electron–positron annihilation the production ratio of hadronic to leptonic states is a quantity that should be little affected by hadronization, at least if one averages over a certain energy interval. For this hadron-to-lepton ratio, called R, different parton models made different predictions. All predict that the ratio is independent of the energy, but the value of the ratio R depends on the model. If one assumes the partons to be quarks of the old quark model with fractional charges and no color, the ratio is the sum of the squared charges of the u, d, and s quarks, that is, $R = (\frac{2}{3})^2 + (-\frac{1}{3})^2 + (-\frac{1}{3})^2 = \frac{2}{3}$.

This result in the old quark model is not only theoretically unfounded, but also much too small. In QCD, asymptotic freedom predicts small coupling at small distances, and the simple calculation is justified. Furthermore, since each quark occurs in three colors, the ratio is larger by a factor of three than that with uncolored quarks, and one obtains $R \approx 2$. Quantum corrections even increase this value a bit. Not only was this calculation theoretically justified, it yielded much better results than the old parton calculations. Nevertheless, there was also a drop of bitterness in this pleasing result. Experimental results might have been somewhat inconsistent in the beginning, but they agreed that the ratio R increased with energy, whereas asymptotic freedom predicted instead a small decrease.

But this seeming discrepancy between theory and experiment found an explanation. In 1974, S. C. C. Ting, at the synchrotron in Brookhaven, and B. Richter, at SLAC, made a discovery that finally accelerated the acceptance of QCD and the standard model. Both groups discovered a new meson with a mass of of 3.097 GeV/c^2. It had only a small mass uncertainty, in contrast to the expectation for a hadron of so large a mass. Its width, 0.09 MeV/c^2, was about 1,000 times smaller than that of the much lighter rho meson. The group at Brookhaven called the new particle J, while the group at Stanford, not without influence from the psychedelic phenomena then in vogue in California, gave it the name psi (ψ). Today, the official name is J/psi (J/ψ).

Not only was the name of the meson controversial at first, but also its nature. One possible explanation was that it was a state that did not transform as a singlet under color transformations. In this case, the color symmetry would have been broken, with disastrous consequences for the renormalizability of the theory. A theoretically more pleasing alternative was to assume that the new meson does not consist of known quarks, but of a new one and its antiparticle. This new quark could very well be the c quark that was proposed by Glashow, Iliopoulos, and Maiani to charm away the problem of strangeness-changing neutral currents.

As we mentioned in Section 6.3, the predicted mass of the c quark was 2 GeV/c^2 or less, which fits in well with the mass of the newly detected J/psi of 3.1 GeV/c^2, that is, less than 4 GeV/c^2.

This interpretation was fully accepted when a year later, after intensive search, a new kind of meson of a mass about 1.8 GeV/c^2 was detected; it was apparently stable under strong interactions and had a "long" mean lifetime of one-thousandth of a nanosecond. This new meson, called the D meson, consists of a c quark and a "normal" antiquark, such as an anti-u quark. The c quark transfers its charm property to the meson, and therefore it cannot decay strongly if it is the lightest meson with this property. The history of the strange quarks repeated itself, but at a much faster pace. In the late 1940s, it was not understood why the "strange" particles lived such a "long" time. Understanding of this concept came in 1953, when Gell-Mann found the explanation in a new property that was called strangeness; see Section 2.2.

In 1971, a Japanese group had reported on "the possible decay of a new particle in flight." It is very probable that this "new particle" had the charm property. But before the discovery of neutral currents and only one year after the speculations about a fourth quark, theorists were apparently not yet ready to wax enthusiastic about "possible new particles."

The small mass width of the J/psi meson could be explained by the OZI rule (see Section 4.3), which states that the quarks contained in a meson must also appear in its decay products; otherwise, the decay is suppressed. According to this rule, the J/psi meson should decay into two D mesons (a D and an anti-D meson, to be precise). However, this is impossible, since the mass of two

D mesons is larger than the mass of the J/psi meson. Therefore, the c quarks in the J/psi meson can only annihilate into virtual gluons, which transform themselves into hadrons.

We have seen that the explanation of the small width of the phi meson, consisting of two s quarks, was instrumental in establishing the quark model. Here, this happened once again, but one can go one step further. The OZI rule is only an empirical rule, but for the J/psi decay, the interaction is of so short a range that the decay rate into (virtual) gluons can be calculated, at least roughly.

Shortly after the discovery of the J/psi meson, another particle that decayed predominantly into the J/psi meson and other hadrons, such as pi mesons, was discovered. This meson, called the psi' meson, consists of a c quark and its antiparticle. It is not in the state of lowest energy, but in an excited state, analogous to the first excited state of a hydrogen atom.

This discovery opened a new branch of QCD, called nonrelativistic QCD. In the limit of very heavy quarks, one can apply nonrelativistic quantum mechanics and can calculate the energies of the excited states using the Schrödinger equation. The interaction at large distances was unknown; for that, one had to make assumptions and determine the free parameters by comparison with experiment. The short-range interaction, on the other hand, can be derived from QCD, and these short-range interactions lead to a characteristic fine structure of the excited states. The overall structure, though not the details, is similar to that of the excited states of the hydrogen atom.

Both theories, QED and QCD, have indeed many common features, and at small distances the two theories differ mainly in the strength of the interaction. To calculate the excited states of a c and an anti-c quark, one could rely on formulas known from atomic physics, and with a bit of knowledge of group theory, one could easily derive the corresponding expressions for QCD. Indeed, one can go beyond nonrelativistic quantum mechanics and construct a full-fledged nonrelativistic quantum field theory—nonrelativistic quantum chromodynamics (NRQCD)— which leads to very good agreement between theory and experiment for heavy-quark calculations.

6.7 Quantum Chromodynamics on the Lattice

The perturbation theory of QCD is impressive as a theory, and it has achieved remarkable success in some domains. But it is not possible to use it to calculate typical properties of hadrons, such as their masses. For that, one would have to know the interaction at distances as large as a hadron, and there, perturbation theory breaks down. The most interesting problem also remains unsolved: why do the field quanta of the elementary fields—the quarks—not show up? Or put more cautiously, why is it the case that free quarks have never been observed?

Therefore, the statement "hadrons consist of quarks" has a completely different meaning from "atoms consist of a nucleus and electrons" or "a nucleus consists of neutrons and protons." An atom can be separated into free electrons and a nucleus, and a nucleus into protons and neutrons. The quarks, however, are apparently confined inside the hadrons. There is a price on the head of the confinement problem: whoever solves it rigorously will receive a prize of one million US dollars.

To me, this seems to be the most exciting question in the physics of elementary particles as it introduced a completely new concept. Just think of the time when Lavoisier revolutionized chemistry by claiming that water is composed of hydrogen and oxygen. If it had not been possible to separate water directly into these elements, I doubt that the "new chemistry" would have found many adherents.

It is therefore very comforting that one can treat QCD at large distances, at least numerically, and can actually calculate such properties as hadron masses. This is done with *lattice regularization.* The first step in this procedure is to assign the quantum fields not to each space-time point but only to a discrete subset, the lattice. Such a lattice in the (two-dimensional) plane is displayed in Figure 6.10(a). It consists of the intersections of the lattice lines, indicated by dots. Of course, the lattice in QCD has not two dimensions, but four—three for space and one for time—but this is difficult to display.

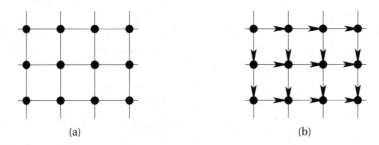

(a)   (b)

Figure 6.10. A lattice in the plane. (a) The dots are lattice points on which the fields are defined. (b) Lattice points and their links. In lattice QCD, the points are assigned to quark fields, the links (arrows) to the gauge fields.

In the early days of quantum field theory one was already assigning fields to lattice points. But if the quark and the gauge fields are assigned to points, an essential property of the theory is lost, namely, gauge invariance. The gauge fields tell us how to compare the properties of fields at different points. In a lattice theory, the points are completely isolated, and therefore gauge invariance is lost.

The solution for a gauge-invariant lattice theory was motivated by ideas from solid-state physics. Franz Wegner had discovered in a model that it is advantageous to use not only the points, but also the links between the points (see Figure 6.10) for the assignment of physical properties. K. Wilson constructed a gauge-invariant version of QCD on the lattice by assigning the quark fields to the lattice points and the gauge fields (gluons) to the links between the points.

In this way, the fundamental equations of QCD can be formulated on the lattice without sacrificing gauge invariance. In the limit where the lattice spacing—the distance between the lattice points—goes to zero, the classical expressions approach the familiar expressions in continuous space and time. If it were of interest, it would be quite simple to numerically solve problems of classical chromodynamics on the lattice, just as it is for problems in electrodynamics. In the latter case, this has technical applications, such as for calculating the optimal design of antennas. However, classical chromodynamics is only of limited interest, and the interesting problems involve essentially quantum corrections, that is, annihilation and creation of virtual particles.

A special feature of QCD is that the gluons can interact among themselves directly. If in lattice calculations one takes only the creation and annihilation of virtual gluons into account, one obtains a very interesting (numerical) result: the interaction energy between two quarks increases as fast as the increasing distance. In order to isolate a quark and an antiquark of a hadron, one has to separate them a (practically) infinite distance from each other, and for that, (practically) infinite energy is needed. This numerical result shows that the confinement property is valid in lattice QCD.

Numerical lattice calculations can be performed only at finite distances. This corresponds to a finite-energy cutoff for the virtual particles in perturbation theory. But in contrast to perturbation theory, the expressions regularized by the lattice method are well defined and not only formal perturbation series. If the lattice spacing is very small, the coupling is very small too, and one can compare the results of a numerical lattice calculation with results of renormalized perturbation theory and see that they are in quite good agreement. This makes it more than plausible that in the limit of vanishing lattice spacing, the lattice results are representative of the theory in the continuum. In general, lattice calculations show good agreement as well with other approximation methods that depend on neither perturbation theory nor lattice regularization.

Since lattice QCD theory starts from the fundamental equations of QCD, only the fundamental parameters of the theory are its input, namely the renormalized coupling strength and the renormalized quark masses. The renormalization scale in this case is the lattice spacing. Though we are far from being able to calculate all hadronic properties, the few that can be calculated show good agreement with experimental values.

An interesting result of lattice calculations is what is known as *deconfinement* at high temperatures. If many quarks are concentrated in a finite volume, one can assign a temperature to them. The higher the average energy of a quark, the higher the temperature. Theoretical calculations on the lattice have shown that at a definite temperature, the interaction between the quarks changes, and where there is a large number of quarks of high average energy, only a finite amount of energy is needed to separate a quark–antiquark pair. One can say that at this temperature, the hadrons melt and form a *quark–gluon plasma*, just as snowflakes melt above zero degrees Celsius and form water.

Despite this deconfinement transition, free quarks cannot be isolated: they cannot be removed from the plasma. According to calculations, this *phase transition* should occur at a temperature of 10^{12} degrees Celsius. This high temperature corresponds to the moderate energy of 100 MeV. It is reported that there are indications for such a melting of nucleons into a quark–gluon plasma in high-energy collisions of heavy atomic nuclei.

6.8   The Consolidation of the Standard Model

The theory of the standard model was completed in 1973. QCD gave a respectable field-theoretic framework for the quark model and strong interactions. With the electroweak field theory, based on the spontaneously broken gauge group $SU(2) \times U(1)$, a renormalizable theory of weak and electromagnetic interaction was found. The importance of asymptotic freedom was fully recognized; the GIM mechanism could explain why there were no neutral strangeness-changing currents; and with the development of lattice gauge theory, one had the possibility to perform serious quantitative calculations outside perturbation theory.

Nevertheless, at that time, the model was not yet standard. In 1973, Martinus Veltman who had contributed so much to the understanding of gauge theories gave a review talk. He said that he would refrain from showing too much enthusiasm, and he gave his talk an epigraph of whimsical German poetry, modeled after the humorous poems of W. Busch, a favorite of A. Einstein. The epigraph stated that there are many theories that can never be tested. This theory, however, can be checked and therefore it will perish miserably. But he then pointed out that it would be difficult to conceal his enthusiasm, "because it seems more and more that we are on the right track." And indeed, on the experimental side as well there were many results that supported the standard model: the predicted neutral currents had been discovered, which allowed a coarse estimate of the mass of the intermediate bosons of the weak interactions. There were also indications for a fourth quark, needed both for the GIM mechanism and the cancellation of anomalies. The following years saw a far-reaching consolidation of the model, but also new challenges, all of which ended triumphantly for the standard model.

An important step was the discovery of *three-jet* events. At the annihilation of highly energetic electrons and positrons there are events in which most particles fly off in two groups, called jets. One can calculate the rate and distribution of these jets in perturbative QCD by assuming that first two quarks are created and that hadronization, that is, the transmutation of quarks into observable hadrons, does not have much effect on the results of the theoretical calculation. There are good theoretical reasons for that, if one concentrates on selected properties of these jet events. Figure 6.11(c) shows the tracks of such a two-jet event, and Figure 6.11(a) displays the simplest graph showing the cause of such an event.

Since in addition to quarks there are gluons, one expects, besides the two-jet events, events with three jets. A typical track configuration and the simplest graphs responsible for these events are shown in Figures 6.11(d) and 6.11(b).

After an intensive search for these three-jet events, several such events were found in 1979 at the electron–positron storage ring PETRA at DESY. They unequivocally pointed to the graphs of Figure 6.11(b) as elementary processes. For this discovery, the prize of the European Physical Society was awarded to P. Söding, B. H. Wiik, G. Wolf, and S. L. Wu. The discovery of genuine three-jet events is referred to, somewhat simplistically, as the discovery of the gluon.

Deep inelastic scattering experiments (see Section 5.8) continued to play an important role in the consolidation of QCD. At the high-energy laboratory DESY in Hamburg, a new kind of storage ring was constructed, in which protons with an energy of 920 GeV collide with electrons of 30 GeV. In such a machine, energies and momentum transfers can be reached that are far beyond

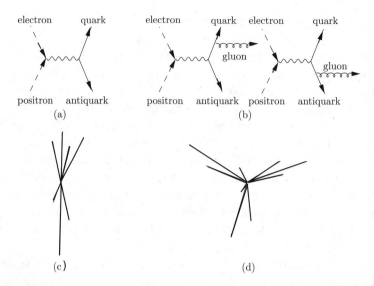

Figure 6.11. Two- and three-jet events in electron–positron annihilation. (a) The simplest graph for two-jet events. (b) The simplest graphs for three-jet events. Dashed lines represent electrons, wavy lines photons, solid lines quarks, and coiled lines gluons. Quarks and gluons are treated here as if they were real particles. (c) The observed tracks of two-jet events. (d) The observed tracks of three-jet events.

the range of those of accelerators with protons at rest. Therefore, deep inelastic scattering could be extended to completely new regions of energy and momentum transfer.

The maximal momentum transfer in the new machine, called HERA, is 200 GeV/c. This corresponds to a resolution of 0.001 femtometer, less than one hundredth of a proton radius. The high energies of the hadronic final state make it possible to measure deep inelastic scattering down to values of the Bjorken scaling variable $x_B \approx Q^2 c^2 / W^2 = 0.00006$. At very high energies and at large momentum transfer, one finds large deviations from the scaling law. These deviations from the results of the naive parton model are expected from QCD and can be quantitatively calculated. Figure 6.12 shows the structure function, which is essentially the form factor, for deep inelastic scattering at two different values of momentum transfer Q. For those values of x_B accessible at the linear accelerator at SLAC, namely x_B larger than about 0.1, the results for the different values of Q are rather similar. The structure function depends only weakly on the momentum transfer Q and strongly on the scaling variable x_B.; that is, it shows scaling behavior.

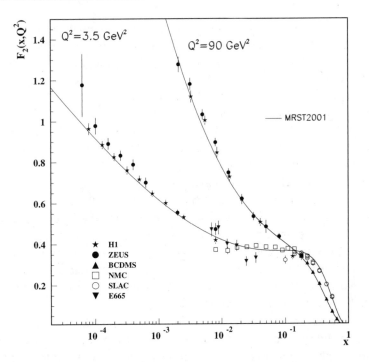

Figure 6.12. The structure function (form factor) of the proton at two different values of the momentum transfer Q as a function of the Bjorken scaling variable $x_B \approx Q^2 c^2 / W^2$. For fixed values of x_B larger than 0.1, the structure function varies little with momentum transfer: it shows scaling behavior. For very small values of x_B, however, the structure function at fixed x_B increases strongly with increasing momentum transfer. This can be interpreted that with higher resolution, one "sees" more gluons inside the proton. The open circles are the results from the electron linear accelerator at SLAC; the filled circles and stars are results from the electron–proton collider HERA.

If one looks into the proton with high resolution (high momentum transfer), the structure function at fixed small values of x_B increases with momentum transfer; that is, scaling behavior is badly violated. But QCD can explain this Q dependence. The solid line for $Q^2 = 90 \text{ GeV}^2$ is, at least in principle, determined from QCD with parameters taken from the curve for $Q^2 = 3.5 \text{ GeV}^2$. Since the variable x_B describes the momentum dependence of the partons, the increase of the structure function with Q shows that with high resolution (large Q), one sees more and more particles with small momentum (small x_B).

The standard model had its most spectacular successes in the electroweak sector. I mentioned earlier that the Nobel committee took a risk in awarding the prize to Glashow, Salam, and Weinberg before the existence of the gauge bosons had been established experimentally. In 1982, the investigations of neutral currents were already at such a high level that one could determine the electroweak mixing angle rather precisely and therefore could make reliable predictions for the masses of the gauge bosons: 77.9 ± 1.7 GeV/c^2 for the charged intermediate bosons W^+ and W^-; 88.8 ± 1.4 GeV/c^2 for the uncharged Z^0. This knowledge made it possible to plan an experiment to find the particles. Carlo Rubbia and his coworkers undertook the task at the proton–antiproton storage ring at CERN.

In this storage ring, a machine of diameter greater than two kilometers, protons and antiprotons collide with an energy from 300 to 400 GeV. The total energy is therefore more than sufficient to create the predicted intermediate bosons. Protons—hydrogen nuclei—are abundant, but antiprotons first have to be created in high-energy reactions and then have to be stored. When enough antiprotons have been collected in a smaller storage ring, they are fed into the big storage ring so that they may collide with the protons. At this moment, they must have a well-defined energy.

For protons, this is no problem: so many are available that one can select just those with the right energy. Antiprotons, however, are precious, and one cannot simply dispose of those without the right energy. To address this concern, Simon van der Meer thought of an ingenious trick, called *stochastic cooling*. The energy of each antiproton is determined. If it is too fast, it is decelerated, while if it is too slow, it is accelerated. With this procedure one can collect a sufficient number of monoenergetic antiprotons. This sounds like a simple idea, but it is as complicated as to teach a bunch of fleas to march in step.

The results of the experiment were worth its extravagance. In early 1983, a paper from Rubbia's group appeared with the title "Experimental Evidence of Lepton Pairs of Invariant Mass Around 95 GeV/c^2 at the CERN SPS Collider." This was the experimental discovery of the uncharged boson Z^0, which decays into lepton–antilepton pairs. Shortly thereafter, the existence of the charged gauge bosons (W^+, W^-) with a mass of 81 ± 3 GeV/c^2 was also established. The present-day values (as of 2007) are 91.1876 ± 0.0021 GeV/c^2 for the Z^0 and 80.403 ± 0.039 GeV/c^2 for the charged gauge bosons W^+ and W^-. The evidence for the particles was clear. Since there are no other particles with similar masses, there was little background noise from other events.

The next challenge soon followed. We have seen that not only was the fourth quark, the c quark, needed in order to charm away the strangeness changing neutral currents, it was also needed to avoid symmetry breaking by

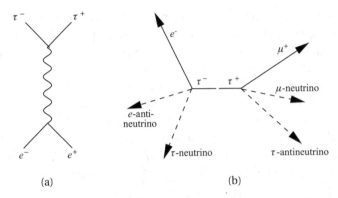

(a)   (b)

Figure 6.13. Production and decay of tau leptons. (a) Feynman graph for the production of tau leptons. (b) Possible decay of a pair of tau leptons into an electron and a positively charged muon. Only the trajectories of the electron and muon are visible; the tau leptons are too short-lived to make visible traces, and the neutrinos do not ionize.

quantum corrections (an anomaly), which would have jeopardized the renormalizability of the weak sector of the standard model. But it was a longstanding question whether there are only two kinds of leptons, electrons and muons.

For the muon, I. Rabi already had posed the question, "Who ordered that?" If it is unclear why there are two charged leptons, there is no reason to believe that there should be *only* two. The first positive indications for a third charged lepton were found in 1974 by M. L. Perl and his coworkers at the electron–positron storage ring SPEAR at SLAC. There were reactions in which both a muon and an electron with opposite charges appeared in the final state. A possible explanation for those events was that the electrons and positrons annihilated and produced a pair of previously unknown leptons, later called tau leptons; see Figure 6.13(a). The tau leptons can decay into an electron, a muon, and neutrinos, as shown in Figure 6.13(b). The heavy tau leptons have too short a lifetime to produce visible traces, and the neutrinos can scarcely be detected.

The number of these events grew, and in 1975, Perl and his coworkers published a paper in which they concluded that the decays cannot be explained by production and decay of known particles, but could be explained very well by the pair production of new leptons with a mass between 1.6 and 2 GeV/c^2. After initial doubts, this interpretation was generally accepted.

The knowledge that pairs of tau leptons can be produced in electron–positron annihilation helped to bring the experimental ratio of hadron-to-lepton production nearer to the theoretical value. This was good news for QCD, but in the electroweak theory, one was in a similar theoretical dilemma as had been the case before the prediction of the c quark. The tau lepton again generates an anomaly, and there are no quarks to compensate for it. But then, as Leon Lederman wrote, nature got terrified: "We found the muon neutrino but missed neutral currents. We discovered what became known as the Drell–Yan process but missed the J/psi. We missed the J/psi again at the ISR but stumbled on high-transverse-momentum hadrons. We missed the J/psi at Fermilab in 1973, chasing single-direct-lepton yields that were a red herring. Then we found a false upsilon. But finally Nature, terrified that she would be stuck with us forever, yielded up her secret, the true upsilon (Υ), hoping this would make us go away."

This upsilon, discovered in 1975, was about three times as massive as the J/psi, but otherwise, the properties were very similar. Therefore, the obvious interpretation was that it comprises a new quark (the fifth one) and its antiparticle. The new quark was called the b quark, for bottom, analogous to the down quark d; the b quark also has charge $-\frac{1}{3}$, and is thus the "lower" member in a fermion doublet, if the latter exists. The new property of this b quark, corresponding to strangeness and charm for the other quarks, was called "beauty," since "bottomness" is presumably too ugly a word. In 1980, hadrons with *open beauty* were detected, that is, particles consisting of one b quark and other conventional ones. The first hadrons of this kind were mesons, seen in 1980 at the accelerator at Cornell University in Ithaca, New York.

This fifth quark, which has about three times the mass of the c quark, does not fully compensate for the anomaly caused by the tau lepton; for that, a sixth quark is necessary. This quark was named the t quark before its birth, that is, before it was experimentally established. The letter t stands for *top*, since it had to have charge $+\frac{2}{3}$, like the u quark (up), in order to compensate for the anomaly.

If one were optimistic, one could say that one had found indications for three lepton quark families (see Table 6.2): first, the family consisting of the electron and its e neutrino as well as the u and d quarks; second, the muon with its mu neutrino as well as the c and s quarks; and third, an incomplete family consisting of the tau lepton and its neutrino and the bottom quark. The second quark of this family, the t quark, badly needed for anomaly cancellation, was missing.

There was yet another theoretical reason for three families. The deeper reason for weak CP violation, mentioned in Section 2.8, was (and still is) not understood. But by investigating CP violation in the framework of the standard

name	symbol	charge	baryon number	lepton number	mass (MeV/c^2)
electron	e	-1	0	1	0.511
e-neutrino	ν_e	0	0	1	< 0.000003
d quark	d	$-\frac{1}{3}$	$\frac{1}{3}$	0	7^*
u quark	u	$\frac{2}{3}$	$\frac{1}{3}$	0	3^*
muon	μ	-1	0	1	105.7
μ-neutrino	ν_μ	0	0	1	< 0.19
s quark	s	$-\frac{1}{3}$	$\frac{1}{3}$	0	120^*
c quark	c	$\frac{2}{3}$	$\frac{1}{3}$	0	1250^*
τ-lepton	τ	-1	0	1	1777
τ-neutrino	ν_τ	0	0	1	< 18.2
b quark	b	$-\frac{1}{3}$	$\frac{1}{3}$	0	4200^*
t quark	t	$\frac{2}{3}$	$\frac{1}{3}$	0	$172\,500^*$

Table 6.2. The three families of leptons and quarks. All leptons and quarks have spin $\frac{1}{2}$, and the quarks have the same internal parity $+1$. *The quark masses are renormalized in the so-called $\overline{MS}$ scheme. For the u, d, and s quarks, the renormalization scale is 2 GeV/c^2; for the heavy quarks, the mass itself is the scale.

model, M. Kobayashi and T. Maskawa found that this violation is compatible with the standard model if there are at least three families. Their paper, which appeared in 1973, increased further the pressure for hunting the sixth quark. In the beginning, the search for it was very difficult, since one had no idea how heavy it was.

One could only suppose from the arrangement of masses in Table 6.2 that it was very heavy. The u and d quarks are very light, the u quark even a bit lighter than the d quark. The next family is considerably heavier, the s quark having a mass of about 150 MeV/c^2, and the c quark about 10 times as heavy. The b quark has a mass of 4.5 GeV/c^2, so the third family seemed to be distinctly heavier than the second one.

Given the speculative tendencies of physicists, there were many theories about the mass of the t quark, but as it turned out later, these theories were nearer to numerology than to science. Since even with increasing energies one saw no signs of a sixth quark, speculative theories were devised that did not

name	symbol	gauge group	mass (MeV/c^2)	width (MeV/c^2)
photon	γ	$SU(2) \otimes U(1)$	0	0
Z boson	Z^0	$SU(2) \otimes U(1)$	91.19	2.50
W boson	W^+, W^-	$SU(2) \otimes U(1)$	80.40	2.14
gluon	g	$SU(3)$	0^*	0

Table 6.3. The gauge bosons of the standard model. All gauge bosons are vector bosons; that is, they have spin 1 and internal parity -1. *The mass of the gluon is, like the quark masses, not a directly measurable quantity.

need a t quark (the so-called topless theories). But finally, it was possible to make a rather precise prediction from the standard model: the mass of the t quark should be in the range from 155 to 185 GeV/c^2.

Where did these numbers come from? Like all successful predictions they are based not on numerology but on empirical data and hard and solid calculations. The data came from the big storage ring LEP at CERN, where electrons and positrons collide with an energy of 100 GeV each; the maximal energy is thus 200 GeV.

With this accelerator, the predictions of the standard model could be tested with great precision. For example, the mass of the neutral gauge boson Z^0 was determined with an accuracy of 0.02% to be 91.1876 GeV/c^2. For such an accuracy, quantum corrections are relevant; that is, the full subtleties of the renormalization procedure enter the picture. Since the virtual t quark contributes to quantum corrections, its mass has influence on their size. Fortunately, the influence of the mass on these corrections is rather large. In order to accommodate all measurements to the predictions of the standard model, it was necessary to use a value of about 170 GeV/c^2 for the renormalized mass of the t quark.

A rather narrow range for the mass having been established, it was much easier to search specifically for the t quark. The hunt was successfully finished in 1995. In the Tevatron accelerator at the Fermilab near Chicago, protons and antiprotons with an energy of 1,000 GeV each collide. Reactions in which t quarks were produced could be observed. The mass was in the theoretically predicted range: around 173 GeV/c^2.

The theoretical prediction and experimental confirmation of the t quark is comparable to the discovery of the planet Neptune. J. C. Adams, of Great Britain, and U.-J.-J. le Verrier, of France, had predicted a new planet from orbital disturbances of the planet Uranus. They theoretically determined its po-

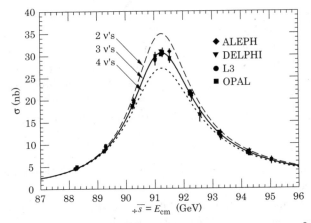

Figure 6.14. The experimental and theoretical resonance curves of the Z^0 boson. The theoretical curve, which assumes three families (three v's) is the only one in agreement with the experimental points from different groups.

sition with fair precision. The prediction was not taken seriously by many observational astronomers, but when le Verrier communicated his results to J.G. Galle at the astronomical observatory in Berlin, the latter, together with his assistant H. L. d'Arrest, found the planet on September 23, 1846, the first night he was searching for it.

With the discovery of the t quark, the three families of leptons and quarks were complete. And what are the prospects for a fourth family? The answer is simple: the prospects are bad! A further result of the precision experiments at the electron–positron storage ring LEP at CERN was that there are exactly three families. Figure 6.14 shows the observed annihilation rate of electrons and positrons into hadrons as a function of energy.

The Z^0 boson appears as a resonance at 91.2 GeV. The width of the unstable "particle" is 2.4952 GeV, from which a mean lifetime of about 2.6×10^{-25} seconds can be inferred. The concept of a particle is here to be understood symbolically. I will return to this problem in Section 8. Let it be noted here that the exact calculations do not depend on the classical particle concept but on an adequate description through a quantized field. Figure 6.14 shows that the theoretical prediction agrees excellently with the experimental data if one assumes that there are three families, more precisely three different kinds of neutrinos. If one assumes that the neutrino of the muon and the tau lepton are identical particles, there are only two neutrinos, and the theoretical curve is far above the data. If one assumes that there are four families with a fourth light neutrino, then the curve is distinctly below the data.

There is one possibility left to accommodate a fourth family with these data: if a neutrino is heavier than 40 GeV/c^2, the Z^0 could not decay into such a neutrino–antineutrino pair, and it has no influence on the curves. The observed neutrinos are all very light. The upper limit for the tau neutrino is the least-stringent one, since there are the least data; the limit is 18 MeV/c^2. Therefore, the restriction to neutrino masses of less than 40,000 MeV/c^2 seems rather mild.

Tables 6.2 and 6.3 collect the known elementary particles of the standard model and some of their properties.

There remains a great challenge for experimental physicists: the establishment of the existence of the Higgs boson, which is responsible for the generation of the masses of the W and Z^0 bosons, the gauge bosons in the weak sector of the standard model; see Section 5.3. Fortunately, the signature of the Higgs boson is very clear. Since it is responsible for mass generation, it couples most strongly to heavy quarks and leptons and decays preferentially into those. Its coupling is strongest to the t quark, but since it is hoped to be lighter than two t quarks, its most probable decay channel is into a b and an anti-b quark. In the observed final state there should therefore appear hadronic states containing these two quarks.

In 2000, there was great excitement at CERN. The electron–positron storage ring LEP was to cease operation at the end of the year in order to make room for the large hadron collider LHC in its tunnel. But shortly before the demise of LEP, the first indications of a Higgs boson with a mass of about 115 GeV/c^2 were thought to have been observed. There was now a difficult decision to be taken: should the operation of LEP be extended and the construction of the large hadron collider LHC delayed? The decision would have not only scientific but also financial consequences; the contracts for the construction of LHC had been signed and there was the problem of high penalties for nonperformance. The main reason for the ultimate decision to shut down LEP was most likely that there was little chance to establish the Higgs boson at LEP beyond reasonable doubt. It may appear strange that I use the phrase "beyond reasonable doubt," an expression that belongs more in a court of law than in a scientific enterprise. Given the statistical nature of measurements in all processes in which quantum physics plays a role, there is always a very small possibility that the "true" result deviates strongly from the measured one, as has been discussed in Section 1.4.3.

Why is the search for the Higgs boson so difficult? One possible reason, of course, is that it does not exist and the mechanism for the generation of the W and Z^0 boson masses is more complicated than has been thought. But even if the Higgs boson exists, as the overwhelming majority of particle physicists think, it is difficult to find. Unfortunately, there is no precise mass range in which to look for it. In the search for the t quark, theory was very helpful. The top mass could be determined from quantum corrections with reasonable accuracy. The Higgs boson contributes to quantum correction, too, but unfortunately, its influence is much weaker than that of the t quark. This is a consequence of the Higgs boson's vanishing spin. Therefore, the theoretical limits for the Higgs mass are rather weak. With 70% probability, it should have a mass between 40 and 203 GeV/c^2; values below 114 GeV/c^2 are experimentally excluded, but that leaves plenty of room in which to search.

6.9   Quark Masses and Their Consequences

The masses of quarks cannot be measured directly, since quarks cannot be investigated as isolated particles. Rather, the quark masses are parameters in the fundamental equations of QCD and have to be determined by comparing theoretical calculations with experimental data. One can calculate masses of hadrons and adjust the quark masses, which are input parameters, to obtain the experimental result. For that, one needs a nonperturbative method such as QCD on the lattice (Section 6.7). Then one can obtain renormalized masses that depend on the renormalization scale; in lattice QCD this scale is determined by the distance of the lattice points. As a reference scale, one usually uses a value of 0.1 femtometer (2 GeV) for the u, d, and s quarks, while for the heavy quarks (c, b, and t) the scale is determined by the mass itself. The quark masses indicated in Table 6.2 are defined in that way.

Since the fundamental equations of QCD can be derived from a so-called *Lagrangian density* (a legacy of classical Euler–Lagrange field theory), quark masses are sometimes called Lagrangian masses. Another name is *current masses*, a reminder of the important role current algebra has played in the development of the quark model. These names were invented to distinguish these masses from the rather vaguely determined *constituent masses*, which will be discussed later. The renormalization of the quark masses is quite delicate, and for different problems, different renormalization schemes and scales must be chosen.

hadron	I	J	P	mass (MeV/c^2)	quark content	sum (MeV/c^2)
π^+-meson	1	0	−	139.6	$u\bar{d}$	10
K^+-meson	$\frac{1}{2}$	0	−	493.7	$u\bar{s}$	123
ρ^+-meson	1	1	−	771.1	$u\bar{d}$	10
a_1-meson	1	1	+	1230	$u\bar{d}$	10
ϕ-meson	0	1	−	1019.5	$s\bar{s}$	240
J/ψ-meson	0	1	−	3096.9	$c\bar{c}$	2400
Υ	0	1	−	9640.3	$b\bar{b}$	8400
neutron	$\frac{1}{2}$	$\frac{1}{2}$	+	939.6	udd	17
proton	$\frac{1}{2}$	$\frac{1}{2}$	+	938.3	uud	13
N^* resonance	$\frac{1}{2}$	$\frac{1}{2}$	−	1535	uud	13
Λ-hyperon	0	$\frac{1}{2}$	+	1115.7	uds	130
Ξ^0-hyperon	$\frac{1}{2}$	$\frac{1}{2}$	+	1314.8	uss	243

Table 6.4. Masses of hadrons, their quark content, and the share of quark masses. Here I is the isospin, J the spin, and P the internal parity. The "mass" column contains the masses of the hadrons, and the "sum" column contains the sums of the quark masses as given in Table 6.2.

The relation between masses of quarks and masses of hadrons is complicated but unequivocal. The quark masses are, apart from the strength of the coupling (color charge), the only free parameters in QCD. The coupling strength is equal for all quarks. This is a consequence of gauge invariance, and therefore the mass differences of hadrons with otherwise equal properties such as spin and isospin must be due to the different quark masses.

It can be seen from Table 6.4 that the sums of the quark masses are in general very different from the hadron mass, especially for light quarks. Nevertheless, there is a remarkable pattern: the differences of various hadron masses are of the same order of magnitude as the differences of the corresponding quark masses. If the masses of the light quarks u, d, and s were all equal, then theoretically the masses of the corresponding hadrons, such as pi and K mesons, neutrons, protons, and lambda and xi hyperons, would all be equal. The flavor SU(3) symmetry (Section 4.2) is therefore a direct consequence of the similarity in the masses of the u, d, and s quarks. This is not astonishing, since we have seen in Section 4.3 that the flavor SU(3) symmetry gave essential motivation for establishing the quark model. The difference between the masses of the d and

u quarks is much smaller than the mass of the s quark. This explains why the isospin SU(2) symmetry is much less broken than the flavor SU(3) symmetry.

If the masses of the u and d quarks were exactly zero, then chiral symmetry would be precisely valid. This means that, independently of each other, the right- and left-handed quarks could be transformed according to the isospin SU(2) symmetry. This is the so-called chiral symmetry $SU(2)_R \otimes SU(2)_L$, where the indices R and L stand for right-handed and left-handed. If chiral symmetry was exact, there should exist another hadron with the same spin and opposite internal parity for each hadron consisting of u and d quarks. The u and d masses are not exactly zero, and therefore we expect a certain breaking of the chiral symmetry on the order of the mass of the u or d quark, that is, around 10 MeV/c^2. There are such parity partners. In Table 6.4 they are listed below the rho meson and the nucleons. However, the mass difference between the partners is not the expected 10 MeV/c^2, but about 50 times that number.

Since we believe that we know the dynamics of strong interactions, there is only one way to explain this huge discrepancy between expected and observed mass difference: in addition to direct breaking of the symmetry by the small quark masses, there must also exist a rather strong spontaneous symmetry breaking (Section 5.3). In the limit of vanishing quark masses, the fundamental equations respect the chiral symmetry, that is, the strict separation into right- and left-handed quarks, but even in this limit, chiral symmetry is broken by the ground state of QCD, the QCD vacuum. In weak interactions, symmetry is broken through an additional quantum field, the Higgs boson. In QCD, the symmetry is broken dynamically: the strong forces give rise to strongly bound quark–antiquark pairs in the vacuum that violate chiral symmetry. One might say that a right-handed quark plunges into the vacuum and emerges as a left-handed quark.

If a symmetry is spontaneously broken, then Goldstone bosons appear. The properties of these bosons are determined by the way the symmetry is broken. In our case, the Goldstone bosons have isospin 1, spin 0, and negative internal parity. The pi mesons have these properties, but they are not massless. This is to be expected for the following reason: the u and d quarks are light, but not exactly massless. The chiral symmetry is therefore not only spontaneously, but also directly, broken; the bosons thereby acquire a mass. The pi meson's mass is determined not only by the mass of the quarks but also by the strength of the symmetry breaking, in our case the abundance of quark–antiquark pairs in the vacuum.

I will not dwell further on the subtleties of chiral symmetry and its breaking. Chiral perturbation theory, which constitutes an entire branch of hadron physics, can be used to calculate a number of hadronic properties.

In the framework of QCD, there should exist not only hadrons made up of quarks, but also those consisting of gluons. These are called *glueballs*. There is an intensive search underway for these particles, and there are some prime glueball candidates. Since a gluon can always transform into a quark–antiquark pair, one cannot expect that glueballs consist exclusively of gluons. There will always be a more-or-less important mixture of quark–antiquark pairs. Thus it is sometimes a question of interpretation whether a certain state is designated as a glueball or simply an ordinary meson.

From time to time, a report on "exotic" hadrons causes a mini-sensation. The last such rumors were about *pentaquarks*, which were very much in fashion during the 2004–2005 particle-hunting season. Those were baryons alleged to be composed of four quarks and an antiquark. In the past, such discussions quickly subsided, since in experiments comprising large numbers of events in which statistical error is consequently small, the occurrence of those exotic states was not confirmed. Even if such states exist, they are not truly exotic, since their existence is by no means excluded by QCD.

I mentioned earlier that besides the well-defined quark masses discussed above—the Lagrangian or current masses—a constituent mass also had been introduced. This mass occurs in simple quark models and is an ill-defined, yet rather useful, concept. The constituent mass of the u and d quarks is somewhere around 300 MeV/c^2, while that of the strange quark is around 500 MeV/c^2. If this mass had been used in Table 6.4 for the sum of the quark masses, it would be much closer to the actual hadron masses. One can imagine that the constituent quark mass includes, in addition to the Lagrangian mass, many nonperturbative effects. For heavy quarks, the constituent mass is near the Lagrangian mass and can even be defined in a theoretically satisfying way as what is called the *pole mass*. One can ascribe a constituent mass to even the gluons. From the few properties of the glueballs that we think we know, the constituent mass of the gluons should be several hundred MeV/c^2.

For completeness, I mention that hadrons consisting of a light and a heavy quark cannot be treated in the framework of perturbation theory. However, with the large mass of the heavy quark there appears a large energy, and thus a short distance scale that entails a certain symmetry. It is described and exploited in the framework of *heavy quark effective theory* (HQET).

In the next chapter, we will return to the "black clouds" on the horizon of the standard model, which can also be regarded—depending on one's level of optimism—as the rosy-fingered dawn of a new physics. But first I would like to outline the standard model—which up to now we have seen only in bits and pieces—in all its glory.

6.10 The Standard Model in All Its Glory

In a letter to Hermann Weyl, Albert Einstein describes how God would have created the world if God had followed Weyl's suggestions. Einstein wanted to convince Weyl that Weyl's original considerations about gauge invariance were elegant but not realized in nature. Einstein concludes, "But since God realized even before the development of theoretical physics that he cannot do justice to the opinions of mankind, he does things just as he likes."

I will use the same procedure in order to demonstrate the consistency of the standard model, if one takes the postulates of gauge invariance and its consequences for quantum field theory seriously. On the other hand, not all theorists' wishes have been satisfied, and there is no justice done to the "opinions of mankind."

Let us start with a particularly simple world with only a U(1) symmetry. Since the gauge bosons of such an abstract symmetry have the properties of photons, we understand the existence of photons, but of nothing else. That is not too bad. After all, photons are by far the most abundant particles of the (known) universe. For every hadron or lepton there are 40 billion (4×10^{10}) photons. Such a universe would be rather dull, since photons do not interact with one another. It therefore seems natural to introduce the states on which the gauge symmetry acts. The mathematically simplest way to proceed is to introduce a massless fermion, described by a Weyl spinor. There are two kinds: right-handed and left-handed. In order to be impartial, we add both. We then have a world of massless charged particles, "electrons" that interact via photons.

We now find ourselves in a universe in which QED is valid. But this theory has a serious shortcoming, since it is not asymptotically free. As mentioned in Section 6.5, such a theory has problems with internal consistency. In order to construct an asymptotically free theory, the gauge bosons have to interact among each other, and therefore the gauge symmetry must be more complex than the U(1) symmetry of electrodynamics. The simplest group that leads to an asymptotically free theory is SU(2), and the simplest objects transforming nontrivially under this group have two elements.

Therefore, we need another particle that forms a doublet with the electron. This is the neutrino. We thereby introduce weak interactions and are a bit closer to reality. In the real world, something happens that we do not understand: left is preferred over right. Indeed, the electron in the doublet is always left-handed, that is, its spin points in the direction opposite to its direction of flight. Moreover, it would be much nicer if the electromagnetic and weak interactions were truly united, but they are not. The two interactions do in fact mix, but we still have two parameters: the coupling strength and a mixing angle. These are so far the only parameters of the theory. Due to this incomplete unification, the problem of asymptotic freedom is also not fully solved, but there are indications of a solution, which will be discussed in the next chapter.

As long as we are criticizing the creation of the universe, let us continue: the gauge bosons of the weak interactions and the electrons acquire mass. This mass does not enter the fundamental equations directly. Instead, it is related to the existence of a further particle, the Higgs boson. The interaction of this boson is such that in the ground state (vacuum), the field does not have the value zero but a finite negative value. This in turn yields a mass for the gauge bosons and electrons. (We shall, for the moment, even though we know better, assume that the neutrinos are massless.)

Now we have five parameters in the theory: besides the mass and the self-coupling of the Higgs boson, there is also a coupling to the electron that yields the mass of the latter; to these three are added the electric and weak coupling, introduced earlier.

The introduction of the weak interaction created a new problem. Its gauge symmetry is broken by a quantum correction, an anomaly, that makes the theory unrenormalizable. Therefore, we have to introduce new particles to compensate for this anomaly. This can be done by introducing fractionally charged quarks with SU(3) as gauge symmetry. This introduces three more parameters: the coupling strength, which is necessary according to theory; and the masses of the two quarks. These masses seem superfluous, but they are there. The matter quantum fields we have now are those of the electron, the neutrino, the u quark, and the d quark. The interactions are mediated by the gauge bosons: the electromagnetic by the photon; the weak by the W^+, W^-, and Z^0; and the strong by the gluons.

As far as we understand, nothing speaks against a universe consisting only of this family. All familiar matter consists of these particles, aside from the

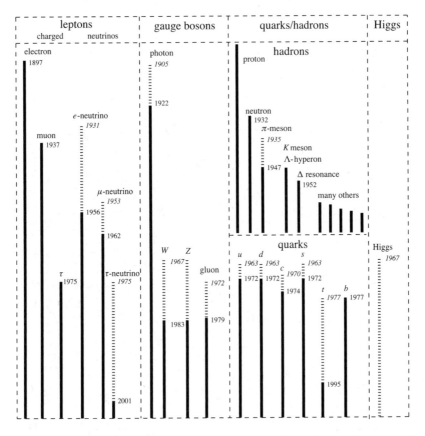

Figure 6.15. The particle zoo. Indicated are the years of theoretical prediction (beginning of dashed lines) and of experimental verification (beginning of solid lines). The dates are sometimes not sharply defined; as stated here, they may be a bit arbitrary. Since about 1970, the hadrons, that is, the observed strongly interacting particles, are no longer considered elementary.

muons and strange particles occasionally appearing in cosmic rays. But life is more complicated than this, and it seems that someone ordered two additional families (Table 6.2). This increases the number of parameters by six (the quark and charged-lepton masses).

   To top things off, the quarks of the different families mix; that is, for the strong and weak interactions, different combinations are relevant. Not only does this seem to be an unnecessary complication; it also increases the number of parameters by four. On the other hand, there might be a deeper reason for

name	symbol	mass (GeV/c^2)	width/ lifetime	J	I	P	S	year
baryons								
proton	p	0.938	stable(?)	1/2	1/2	+	0	
neutron	n	0.939	886 sec	1/2	1/2	+	0	1932
Λ-hyperon	Λ	1.115	2.6×10^{-10} sec	1/2	0	+	-1	1951
Σ-hyperon	Σ^+	1.189	0.8×10^{-10} sec	1/2	1	+	-1	1953
	Σ^0	1.193	7.4×10^{-20} sec	1/2	1	+	-1	1956
	Σ^-	1.197	1.5×10^{-10} sec	1/2	1	+	-1	1953
Ξ-hyperon	Ξ^0	1.315	2.9×10^{-10} sec	1/2	1/2	+	-2	1959
	Ξ^-	1.321	1.6×10^{-10} sec	1/2	1/2	+	-2	1952
Δ resonance	Δ	1.232	120 MeV	3/2	3/2	+	0	1952
Σ^* resonance	Σ^*	1.385	37 MeV	3/2	1	+	-1	1960
Ξ^* resonance	Ξ^*	1.533	9.5 MeV	3/2	1/2	+	-2	1962
Ω	Ω	1.672	0.8×10^{-10} sec	3/2	0	+	-3	1964
Y^* resonance	Y^*	1.406	50 MeV	1/2	0	−	-1	1961
mesons								
π-meson	$\pi^\pm$	0.140	2.6×10^{-8} sec	0	1	−	0	1947
	π^0	0.135	8.4×10^{-17} sec	0	1	−	0	1949
K meson	$K^\pm$	0.494	1.2×10^{-8} sec	0	1/2	−	±1	1951
	K^0_S	0.498	0.89×10^{-10} sec	0	1/2	−	*	1954
	K^0_L	0.498	5.2×10^{-8} sec	0	1/2	−	*	1954
ρ-meson	ρ	0.771	150 MeV	1	1	−	0	1961
ω-meson	ω	0.782	8.4 MeV	1	0	−	0	1961
η-meson	η	0.547	0.0012 MeV	0	0	−	0	1961
ϕ-meson	ϕ	1.02	4.6 MeV	1	0	−	0	1962
J/ψ meson	J/ψ	3.096	0.087 MeV	1	0	−	0	1974
ψ' meson	ψ'	3.685	0.3 MeV	1	0	−	0	1974
D meson	$D^\pm$	1.869	1.0×10^{-12} sec	0	0	−	c	1975
	D^0	1.865	0.4×10^{-12} sec	0	0	−	c	1975
Υ	Υ	9.460	0.052 MeV	1	0	−	0	1974
B meson	$B^\pm$	5.279	1.7×10^{-12} sec	0	0	−	b	1980
	B^0	5.279	1.5×10^{-12} sec	0	0	−	b	1980

Table 6.5. The hadrons mentioned in the book. Under "width/lifetime," the mean lifetime in seconds is indicated if the particle is stable under strong interactions; for unstable particles, the width is registered. Here J indicates spin, I isospin, P internal parity, and S strangeness, charm (c), or beauty (b). Here "year" indicates the year of experimental verification; frequently, there is a long period between first hints of a particle and its final identification. Therefore, the dates are only approximate.

*The K^0_S and K^0_L have no defined strangeness as superpositions of K^0 and $\bar{K}^0$.

three families: only for three families can a breaking of the symmetry between matter and antimatter (CP-invariance) be made compatible with the standard model. Breaking the symmetry might explain the asymmetry in our universe in which only matter, and not antimatter, occurs naturally. For some time, there was optimism that this asymmetry could be explained by the observed violation of CP-invariance, but this optimism has dwindled.

We see that there is a remarkable consistency, but also see some features of our universe that are less elegant. Many physicists think that the number of free parameters, namely 18, is too high for a fundamental theory. That is, of course, a very immodest claim. Isaac Newton's theory was hailed with enthusiasm, though it needed many more than 18 parameters to describe the solar system.

However, there are indeed several dark clouds hanging over the standard model. Some may be real, some imaginary. These clouds and some ambitious projects to enlarge or even replace the standard model will be treated in the next chapter.

Figure 6.15 summarizes the particle content of the standard model in its historical development. The number of leptons has increased and a new structure has been developed in the second half of the twentieth century. Each charged lepton together with its neutrino forms a doublet. For the hadrons, there was a break around 1970 (a revolution if you like strong words): the hadrons are no longer considered as elementary, but consisting of quarks. To the three doublets of leptons there correspond three doublets of quarks. All interactions—the strong, the electromagnetic, and the weak—are mediated by gauge bosons. The hadrons that have been mentioned in this book are collected in Table 6.5.

7

Storm Clouds or the Dawn of a New Physics?

Most particle physicists hope that the standard model is not exactly correct; they hope for fundamentally new physics, and discrepancies between theory and experiment could be indications of that. In this chapter, I first will give some indications about such deviations; then I will describe how the standard model can be extended in such a way that the new version is valid up to distances ten trillion times smaller than those that can be presently resolved. This would be a giant step in the region beyond the nanoworld. But speculations go even further, as will be briefly discussed in the last section.

7.1   Neutrinos, Too, Are Out of Tune

This section starts with the most important discovery of particle physics in the new millennium, neutrino oscillations, and it ends with some minor "detuning" of the standard model.

I mentioned in Section 2.3 that small mass differences lead to particle oscillations. Over time, one type of particle changes continuously into another and then back again; these oscillations are analogous to the beats of slightly detuned strings. We now have very good evidence that such a phenomenon occurs with neutrinos. For instance, a mu neutrino can transform into a tau neutrino, or an *e* neutrino into a mu neutrino. From this phenomenon, there are two things to be learned: the different families are not separated by strict selection rules, and at least one of the neutrinos must be massive, since oscillations are due to mass differences. It is not yet clear whether this leads to merely a simple extension of the standard model or to important reconstructions of the foundations. Therefore, I will not speculate on the further consequences and confine myself to the exciting history of the discovery of neutrino oscillations.

The search for the mass of the neutrino is as old as the neutrino hypothesis itself. Whenever a new type of particle is introduced, physicists want to know its mass. The search for the neutrino's mass continued even after the discovery of maximal parity violation made it plausible that neutrinos are described by Weyl spinors and therefore are massless.

But one never knows for certain, and so one hedges one's bets. The first attempts to measure the neutrino's mass were made in beta-decay experiments, in which by measuring the momenta and energies of the charged particles one can determine the mass of the invisible neutrino. An indication of a finite neutrino mass was not found, but the errors in measurement meant that only an upper limit could be given.

For the e neutrino, this mass limit is $3 \, \text{eV}/c^2$; for the mu neutrino, the mass limit is $0.18 \, \text{MeV}/c^2$; and for the tau neutrino, the mass limit is $18 \, \text{MeV}/c^2$. The different values are due to the fact that for the tau neutrino, for instance, there are much less data than for the e neutrino, and therefore the errors are larger.

I mentioned earlier that oscillations allow a much more precise determination of masses than direct measurements, just as the tuning of a musical instrument by beats is much more precise than by adjustment of the absolute pitch of the tones. A prerequisite for the existence of beats is, of course, that transitions between the neutrinos from different families be possible. This is quite plausible, since weak interactions also allow transitions between quarks of different families.

The first serious hints of a transition for e neutrinos came from discrepancies between astrophysical calculations and the experimentally observed neutrino flux from the sun. Energy production in the sun can be reduced to the transformation of protons into neutrons by the weak interaction and the formation of atomic nuclei. The first calculations had been made by H. Bethe and C. F. von Weizsäcker in 1937. Later, it turned out that the direct fusion of two protons into a heavy hydrogen nucleus is the first major step in the process. Two protons collide, and one proton is transformed by weak interaction into a neutron, a positron, and an e neutrino. The neutron and the proton form a deuteron, a nucleus of heavy hydrogen; thereby an energy of about 1.5 MeV is released. This process is displayed graphically in Figure 7.1. After further nuclear interactions among other nuclei, a boron isotope is produced consisting of five protons and three neutrons. It decays rapidly into a positron, a neutrino,

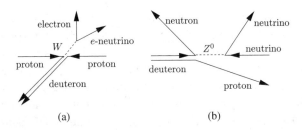

(a)   (b)

Figure 7.1. (a) The heating of the sun. Two protons produce a deuteron (the nucleus of heavy hydrogen), a positron, and an e neutrino. The intermediate state is a virtual W^- boson. (b) Detection process for neutrinos with the help of heavy hydrogen. A neutrino of an arbitrary family splits a deuteron into a proton and a neutron; a virtual Z^0 boson is exchanged.

and an unstable isotope of beryllium. The final product is helium. Highly energetic neutrinos from boron decay are particularly suitable for detection.

The photons produced in nuclear reactions and in positron annihilations heat the sun, and the neutrinos leave the sun practically unhindered and therefore carry energy away and cool the sun. This cooling process is an essential part of all solar models. Therefore, one can calculate the number of produced e neutrinos quite accurately. The number of e neutrinos reaching the earth can be measured in very large-scale experiments. All these experiments are performed deep under the surface of the earth, since the thick layer of matter shields the detectors from cosmic rays, whereas the neutrinos can penetrate it practically without loss. Since the reaction probability of neutrinos is very small, counters with a huge volume are necessary to detect them.

A pioneering experiment was performed by the physical chemist R. Davis and his collaborators in the Homestake Gold Mine, in South Dakota. The mine is 1,500 meters below ground, and the detector was a tank filled with 400,000 liters of perchlorethylene, a liquid that formerly was much used for dry cleaning. The solar neutrinos can transform a neutron in the nucleus of a chlorine atom into a proton, thereby increasing the charge by 1, and the resulting element is the noble gas argon. The reaction equation is $v_e + {}^{37}Cl \rightarrow e + {}^{37}Ar$. The resulting argon is radioactive and can be detected through its decay. The experiment of Davies and his collaborators produced the "neutrino crisis." The number of detected neutrinos was only one-third of what had been expected from model calculations of J. Bahcall. This unexpected result led to further experiments in which solar neutrinos were detected by other processes.

In the Gallex experiment, the detection device was a tank with 54 cubic meters of a solution of gallium chloride in hydrochloric acid. The experiment took place in a laboratory near the road tunnel under the Gran Sasso massif, in Italy. Here an e neutrino transformed gallium into germanium and an electron. The germanium atoms can be separated chemically from the gallium and detected through their radioactive decay. In contrast to the chlorine–argon reaction of the Homestake experiment, even low-energy neutrinos can transform gallium into germanium.

The detection device of the Japanese Super-Kamiokande experiment was a tank of 50,000 tons of purified water 1,000 meters underground. In the experiment, e neutrinos hit the electrons of the water molecules, as displayed in Figure 6.1. The very fast electrons emit Cherenkov light (see Section 1.3) that was registered in 11,200 photocells. Compared to the enormous dimensions of Super-Kamiokande, the gigantic bubble chamber Gargamelle, mentioned in Section 6.3, seems a dwarf. The yield of Super-Kamiokande was enormous: in 300 days, 44,000 solar neutrinos were detected.

In these and in some other experiments the discrepancy between experimental result and theoretical expectation was fully confirmed. There were, of course, still some open questions. Was the theoretical calculation with the solar model really reliable? Were the large-scale experiments as precise as expected? But given the high level of consistency among the different experiments, the only viable conclusion was that either the solar model was wrong or a fraction of the e neutrinos disappeared during their travel from the sun to the earth. For the detection of solar neutrinos, only processes caused by e neutrinos were used. Therefore, a possible explanation for the neutrino deficit was that a fraction of the solar e neutrinos had changed on the way from the sun to the earth into mu or tau neutrinos and thereby escaped detection.

New experiments showed that there was indeed such a conversion. In these experiments, made at the Sudbury Neutrino Observatory (SNO), in Ontario, the fission of heavy hydrogen nuclei by neutrinos was observed. This reaction can be induced by any neutrino, and not only by e neutrinos. It is represented in Figure 7.1(b). The experiment was possible only because in Canada, large amounts of heavy water are available from nuclear reactors. One thousand cubic meters of heavy water were borrowed from the reactor companies, and this water constituted the measuring volume in the experiment, which took place in a mine 2,000 meters below the surface of the earth.

If the solar model is correct and the discrepancy between theoretically calculated and experimentally observed e neutrinos is due to a transformation of

experiment year	detection	neutrino flux
solar model I 2005		5.69 ± 0.9
solar model II 2001		5.31 ± 0.6
Kamiokande 1996	*e* neutrinos	2.8 ± 0.4
Super-Kamiokande 2005	*e* neutrinos	2.35 ± 0.09
SNO 2005	*e* neutrinos	2.35 ± 0.27
	all neutrinos	4.94 ± 0.43

Table 7.1. Model calculations and detection experiments for the flux of solar neutrinos originating from the decay of the boron isotope ^{8}B. The unit of flux is one million neutrinos per square centimeter per second. Kamiokande and Super-Kamiokande are the two Japanese experiments; SNO stands for Sudbury (Canada) Neutrino Observatory.

e neutrinos into mu or tau neutrinos, then there should have been no deficit of neutrinos in the Sudbury experiment, since the transformed neutrinos are also detected.

Table 7.1 presents the newest results. In the Kamiokande experiments, only the *e* neutrinos were detected, with the result of the famous deficit, but in the Sudbury (SNO) experiment, neutrinos of all families were detected with equal efficiency, and there was no discrepancy between the theoretical and experimental numbers. This was to be expected if the solar model is correct and the deficit is due to a transformation of the *e* neutrinos into those of other families. The agreement is perfect, within experimental error.

This highly convincing indication of neutrino oscillation (actually, oscillation in only one direction) is corroborated by two other methodologically independent experiments. In the subterranean experiment Super-Kamiokande, one was also looking for mu neutrinos; they have their origin in the decay of muons from cosmic radiation. They are produced in the atmosphere, and it seems logical that as many mu neutrinos should enter the counters from above as from below. However, this was not the case. There were distinctly more neutrinos coming from above than from below. Neutrino oscillations offer a natural explanation for such a result: the neutrinos coming from below are mostly produced in that part of the atmosphere that is on the opposite side of the earth (that is, over South America). Their path through the earth is much longer than for those produced directly over Japan; therefore, they had much more time to transform into another family. All details are very well explained if one assumes that the mu neutrinos transform into tau neutrinos.

Another compelling argument for neutrino oscillations comes from a reactor experiment in Japan. The neutrinos originating in power reactors are detected in a 1,000-ton liquid scintillator detector at a distance of about 180 kilometers. Only about two-thirds of the neutrinos generated in the reactors were detected at this distance. Accelerator results from Los Alamos are awaiting confirmation.

Since mass differences are the cause of transformations, at least one of the neutrino families must have massive neutrinos. From the experiments, one may conclude that at least one kind of neutrino must have a mass of at least $0.09\,\mathrm{eV}/c^2$.

R. Davis and M. Koshiba were awarded the 2002 Nobel Prize in physics for their "pioneering contributions to astrophysics, in particular for the detection of cosmic neutrinos."

I mentioned earlier that it is not clear to what extent the standard model has to be modified in view of the neutrino masses, but the next few years will certainly bring important information. At the moment, there is an extensive discussion around the question of whether the neutrinos oscillate only inside the three known families or whether there is also a fourth, sterile, neutrino. This sterile neutrino does not couple to the gauge bosons of electroweak interactions, but it could have an important place in an extension of the standard model. Another important question is whether the neutrinos are Majorana or Dirac fermions. For Majorana fermions, particles and antiparticles coincide, but not for Dirac fermions. For massless neutrinos this question is irrelevant, but for massive neutrinos it is important for the extension of the standard model. Normally, one assumes that neutrinos are Majorana fermions, but this has by no means been proved.

Besides the clear-cut effects for neutrinos that certainly make a modification of the standard model necessary, there exist as well some minor discrepancies, the significance of which is uncertain. An example of such a discrepancy is a certain asymmetry in the decay of Z^0 bosons into particles containing b quarks and anti-b quarks. This asymmetry is expressed by an asymmetry parameter $A_{FB}^{0,b}$. The theoretical value is $A_{FB}^{0,b} = 0.0982 \pm 0.0017$, while the experimental value is 0.1036 ± 0.0008.

This discrepancy is just three standard deviations. It is hard to say how seriously it need be taken. Besides the unavoidable statistical fluctuations, one also has to take into account that there is no way to evaluate beyond any doubt the necessarily nonperturbative corrections from the strong interaction. Furthermore, the experiments are sometimes plagued by systematic experimental errors. An example is to be found in R_b, the ratio of Z^0 decays into particles with b and anti-b quarks to all other hadronic decays. Here for some time a

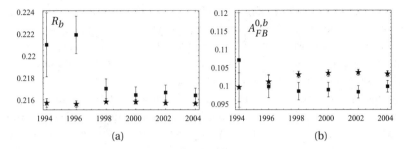

Figure 7.2. Experimental values (squares) and predictions of the standard model (stars) over time. (a) The ratio R_b. (b) The asymmetry parameter $A_{FB}^{0,b}$.

discrepancy of more than three standard deviations existed, but over time, the discrepancy shrank to be insignificant. Figure 7.2 shows the predictions of the standard model and the experimental values for these quantities.

Before we come to more fundamental—one could almost say ideological—objections to the standard model,I will digress from this topic to focus on the problem of particles' masses.

7.2 Why Do Elementary Particles Have Mass?

In the simplest of all possible worlds, all elementary particles would be mass-less. The fermions, that is, leptons and quarks, would be described by Weyl spinors. These obey the chiral symmetry, which excludes a mass. The gauge bosons—that is, the photons, the intermediate $W^{\pm}$ and Z^0 bosons, and the gluons—would all be massless because of the gauge symmetry. But we have seen that through the dynamics of the Higgs boson, the electroweak gauge bosons as well as the fermions acquire a mass. Nevertheless, the statement "all elementary particles are massless" is true to a very good approximation. The masses of elementary particles are very small, almost incredibly small. Thus arises the question, what does "small" mean? That the masses are small in units of everyday life, such as ounces and pounds, is natural. But the masses of the elementary particles are also very small in natural units. These units were introduced by Planck in 1899 and are based on the velocity of light in a vac-uum, on Newton's gravitational constant, and on Planck's constant. They are described in the appendix on units, Appendix B.

The natural unit for mass, the Planck mass, is not large in everyday units, but at two-hundredths of a milligram, it is nearly within the range of apothecary weights. This implies that the Planck mass in units appropriate to elementary particles, that is, GeV/c^2, is very large, namely 1.22×10^{19} GeV/c^2. Compared to that, the mass of the Z^0 boson, with its 90 proton masses a heavyweight among the elementary particles, is infinitely small, as small as the mass of a fly compared to that of 10 million elephants.

In this light we can regard the families of fermions and the gauge bosons as massless to a very good approximation, and the above-mentioned symmetries give us good reason for that. There remains one particle, the Higgs boson. For that particle none of the known symmetries predicts a vanishing mass value, and therefore its natural mass would be on the order of the Planck mass. The Higgs boson has not yet been detected, and one knows only that it has to be heavier than 115 GeV/c^2, if it exists. There are, however, very convincing theoretical arguments that it should not be heavier than 500 GeV/c^2, which is far, far below the Planck mass. There should be an additional symmetry that makes the Higgs boson light. One such symmetry is supersymmetry, discussed in Section 7.4.

In recent years, the possibility of obtaining a smaller value for the natural mass unit has been discussed intensively, based on the fact that general relativity is very well tested for large distances but not for small ones. It is possible that Newton's law, according to which the gravitational force falls off with the square of distance, is not satisfied for very small distances. Such deviations from Newton's law could be due to higher dimensions. In such theories, our three-dimensional space is embedded into a space of higher dimension, which can be penetrated only by gravitational forces. In such a higher-dimensional space, the Planck mass would be much smaller. Two additional space dimensions would reduce it to approximately 100 GeV/c^2, which is exactly the range of electroweak symmetry breaking.

A theory with extra space dimensions was discussed earlier, by T. Kaluza (1921) and O. Klein (1926); they were stimulated by the work of Weyl on gauge symmetry in gravitational theory (see Section 5.1). Since even the most cloistered mathematical physicist has to move in three dimensions of space and one of time, he or she must admit that the extra dimensions cannot be treated on the same footing as the familiar ones. The problem is solved by *compactification* of the extra dimensions. Such a compactification can be viewed intuitively as follows: a sheet of paper is a two-dimensional object. However, if it is rolled up very tightly, one of its dimensions, its length, is still visible macroscopically, while the other dimension, when viewed from a large distance, seems to have disappeared; it has been compactified.

These speculations on extra dimensions may sound bizarre, but they lead to experimentally testable predictions. It might well be that for a future generation of physicists extra dimensions will be as familiar as the violation of symmetry under space reflections is today. We will again venture into extra dimensions in Section 7.7.

7.3 The Grand Unification

The most compelling objection against the standard model in its present form is rather subtle: there are aspects of the model that are not asymptotically free. As described at the end of Section 6.5, theories are presumably inconsistent if the interaction increases with decreasing distance. In the standard model there are two interactions that have this suspect property: the electromagnetic interaction and the self-interaction of the Higgs boson. Whereas there is no remedy in sight for the Higgs interaction, there exists an elegant solution for the electromagnetic interaction, called *grand unification.*

In the standard model, the electromagnetic and weak interactions are not truly unified. The mathematical reason for that is the occurrence of a direct product in the symmetry group SU(2) ⊗ U(1). The physical consequence is that there are two independent coupling constants, one for SU(2) and one for U(1). It would therefore be more appropriate to speak of a mixing of weak and electromagnetic interactions than of a unified electroweak theory.

It turned out that a true unification of the electroweak sector alone is not promising, but a grand unification of the electroweak and strong interactions in one gauge group offers very attractive prospects. In this theory quarks and leptons are transformed by one and the same gauge group, and the gauge bosons of this group describe the electromagnetic, weak, and strong interactions. This is the *grand unified theory* (GUT). As a nonabelian gauge theory, it is asymptotically free. It also helps in removing an unpleasant feature of the standard model, in which the quarks and leptons are completely unconnected, although the combination into families is necessary in order to compensate for the anomalies, as described briefly in Section 6.8.

An important consequence of grand unification is that there is only one coupling constant for all interactions. However, at presently attainable energies (distances), this is by no means the case. This indicates a strong violation of the symmetry. Therefore, the predictions of grand unified theories seem rather useless, yet such is not the case. Though the symmetry might be valid at energies that are not attainable, we can test the consequences with present-day methods. The key to doing so is the renormalization group, which predicts how the couplings change with varying scale (see Section 6.5).

At distances where the unified symmetry is not yet valid, the coupling constants behave according to the predictions of the standard model. The strong coupling and the weak coupling—more precisely, the coupling corresponding to the SU(2) part—fall off with increasing energy, while that corresponding to U(1) increases. At the (high) energy (small distance) where the unified theory is valid, there is only one coupling, and the three must meet at its value. Starting with this energy (unification scale), the coupling decreases with increasing energy (decreasing distance). This means that the theory is asymptotically free if the energy is large enough. The expected behavior is shown in Figure 7.3(b).

The meeting point of the three coupling constants can be theoretically calculated in specific models. The first estimates already had shown that this point is at a distance scale of about 10^{-17} femtometers, in energy units about 10^{16} GeV. This is still smaller than the Planck scale by a factor of 1,000, but far above energies attainable in the foreseeable future. Many physicists were shocked by this high energy, for it signifies that between the presently attainable energies of about 1,000 GeV and this truly astronomical energy of 10^{16} GeV, nature behaves as predicted by the standard model. Therefore, in this huge energy range no new discoveries are to be expected; one spoke of the "great desert."

But this large scale had an advantage for the survival of the theory. It makes a decay that is unavoidable in a grand unified theory, but that has not yet been observed—namely, the decay of the proton—very weak. Since quarks and leptons are in the same representation of the symmetry group in the grand unified theory, this symmetry predicts interactions that transform quarks into leptons. If quarks can transform into leptons, then protons and neutrons in the nuclei can decay into leptons. If the symmetry group is fixed, theory predicts the strength of the interaction and thus the lifetime of the proton.

In order to make quantitative predictions, one therefore has to know the symmetry group. It must contain the symmetry groups of the standard model, that is, SU(3) of the strong interaction and SU(2) ⊗ U(1) of the electroweak interaction. Technically speaking, this means that these groups have to be subgroups of the unified symmetry group. For some time, the group SU(5) had many adherents. For the lifetime of the proton, it yielded about 3×10^{32} years. Compared to this lifetime, the universe is very young. The 13 billion years since the big bang have the same ratio to the calculated lifetime of the proton as 10 microseconds to the age of the universe.

One might at first think that such a long lifetime is a safe bet, since nobody can check it, but fortunately, that is not the case. The lifetime is long, but the number of protons is very large. A gram of water contains about 6×10^{23} protons or neutrons, and so in the tank of the Super-Kamiokande experiment (with which we became acquainted in Section 7.1), with its 50,000 metric tons of water, there are more than 10^{34} nucleons (that is protons or neutrons bound in the oxygen nucleus). With a mean lifetime of 10^{32} years, one thus expects more than 100 decays per year.

The many experiments looking for nucleon decay are all situated deep underground, and the main source for the disturbing background is neutrino reactions. Thus we have learned much about neutrinos, as we saw in Section 7.1, but up to now, no proton decay has been observed. According to the present status of the experiments, the lifetime for the decay of a nucleon into a positron and a pi meson is greater than 5×10^{33} years. With this value, the once very popular symmetry group SU(5) is ruled out. On the one hand, this is a pity, but from a more comprehensive point of view, it is also welcome, for it shows that theoretical speculations on the behavior at scales that are not directly attainable can nevertheless be tested experimentally.

Another source of trouble for the grand unified theory came from the precision experiments performed at the large electron–positron collider LEP at CERN. The coupling constants were determined so precisely that one could very reliably extrapolate them to the large unification scale. It thus turned out that in all realistic unified theories, the three couplings of the strong, electromagnetic, and weak interactions did not meet in one point, as can be seen in Figure 7.3(a). But here a new symmetry helps, called *supersymmetry*. It had been investigated for quite some time, but earlier there had been no experimentally founded indication for it in particle physics. We will discuss it in the next section.

7.4 Supersymmetry

Supersymmetry is a new kind of symmetry, which relates particles of different spin, for instance those of spin 1 with those of spin $\frac{1}{2}$. If supersymmetry was exact, then to each particle with spin $\frac{1}{2}$ there should exist a particle with the same mass and spin 0 or 1. If supersymmetry is not exact, particles with different spin should have approximately the same mass. The hope that the known particles with spin 0, $\frac{1}{2}$, or 1 can be unified into a *supermultiplet* is too optimistic. A nice feature of supersymmetry is that it predicts many new particles. But a bad feature is that none of those particles has yet been observed. No wonder that experimentalists were not very enthusiastic about supersymmetry, no more than Anderson was about Dirac's "esoteric theory" of antiparticles.

Supersymmetry has a somewhat diffuse history. It began to be taken seriously as a theory only after the seminal work of Julius Wess and Bruno Zumino, which appeared in 1974. One year later, R. Haag, M. Sohnius, and J. T. Lopuszanski discovered that a supersymmetric field theory plays a unique role: it finds exactly the loophole that a theory needs to circumvent a theorem of S. Coleman and J. Mandula. Somewhat simplified, this theorem states that no symmetry can exist that relates particles of different spin. This theorem is circumvented in supersymmetry by the introduction of a new kind of generator that does not follow commutation relations (see Section 1.4.2) but instead obeys *anticommutation* relations. These new generators connect particles that differ in spin by one-half, such as those with spin 0 with those of spin $\frac{1}{2}$.

This is not only mathematically remarkable, but also useful for the standard model. We saw in Section 7.2 that there is no satisfactory reason for a light Higgs boson: its natural mass would be on the order of the Planck mass. For fermions, however—that is, for particles with spin $\frac{1}{2}$—there is a reason to be massless, namely, chiral symmetry. Since the Higgs boson is together in the same supermultiplet with fermions, the Higgs boson must also be light. If supersymmetry were exact, the Higgs boson would have the same mass as the fermions, while if supersymmetry is broken, the Higgs mass is of an order of energy that breaks the supersymmetry. Therefore, supersymmetry would explain a light Higgs boson.

These arguments were taken so seriously that not only mathematically minded theoretical physicists, but also those interested in experiments, took a closer look at supersymmetry. They convinced the experimental physicists that it is worthwhile to search for the supersymmetric partners of the known particles. Though not one of the many predicted particles has been found, they already have been given names. The partners of bosons with spin 0 or 1 are fermions. Their names are characterized by the suffix "-ino"; neutralinos are the partners of the Higgs boson and the Z^0 boson, charginos are partners of the charged gauge bosons, while the photino and the gluino correspond to the photon and the gluon. The partners of the fermions have the prefix "s-"; squarks and sleptons, or more specifically selectron, stop, etc.

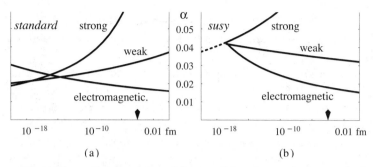

Figure 7.3. (a) The scale dependence of the coupling constants in the standard model. (b) The scale dependence in the supersymmetric standard model.

The precision experiments at LEP provided additional support for super-symmetry. I mentioned at the end of the last section that the grand unified theory has difficulties in the standard model, since the three couplings of strong, weak, and electromagnetic interactions do not meet at one point; see Figure 7.3(a).

Here supersymmetry works wonders: if the standard model is modified to a supersymmetric one, the coupling constants meet at one point, at the scale of 6×10^{-18} femtometers, corresponding to an energy of 3×10^{16} GeV; see Figure 7.3(b). At small distances, the common coupling decreases with decreasing distance: the theory is asymptotically free.

Supersymmetry must be valid not only at this high unification scale, but at energies that are presently just available. In length units, this is 0.0002 femtometer; in energy units, 1,000 GeV. With masses of this order of magnitude, one should be able to produce and detect some supersymmetric partners of known particles, either at the Tevatron at the Fermilab, or at the future Large Hadron Collider (LHC) at CERN.

The lightest supersymmetric particle is presumably a neutral fermion, a neutralino, and the search for it is especially intensive. It interacts only weakly with the known particles, and one could find it by a seeming violation of energy and momentum conservation. In a reaction, the missing energy and momentum are then carried away by an unobserved particle like the neutralino. The situation is similar to that of beta decay, where Pauli used the seeming violation of energy conservation to deduce the existence of a neutral, weakly interacting particle, the neutrino.

Indeed, some high-energy reactions with missing momentum have been observed at Fermilab, but they are not yet statistically significant: with a few percent probability, conventional explanations for the missing momentum are still possible. At present, one can determine only secure lower bounds for the mass of the lightest neutralino, namely 40 GeV/c^2.

In the supersymmetric standard model, the Higgs boson should have a mass of less than 150 GeV/c^2, while the present lower limit for the Higgs mass is 115 GeV/c^2. The lifetime of the proton depends on the special group. From the negative results at Super-Kamiokande, we know that the partial lifetime for the decay of a proton into a positron and a neutral pi meson is larger than 5×10^{33} years. These results already rule out the simplest (minimal) supersymmetric standard model, but more sophisticated models yield values that are about five times larger than the experimental lower limit.

It seems likely that the coming years should force a decision about the fate of the supersymmetric standard model. Either some of the predictions will be confirmed experimentally, or the model will fade away. If the model is corroborated, there will be great motivation for further investigation. Precision experiments at around 1,000 GeV could test effects at the unification scale of 10^{16} GeV.

7.5   Monopoles

The grand unified theory renewed interest in an old problem: whether or not there exist magnetic charges similar to the electric ones. Magnetic charges are called (magnetic) *monopoles*. Classical electrodynamics shows a remarkable correspondence between electric and magnetic phenomena, but also a remarkable difference: there exist electric charges, but no magnetic charges. If we cut a magnetic dipole such as a magnetic rod, we do not separate the magnetic poles, we obtain two dipoles, each of which has a magnetic north pole and a magnetic south pole.

In classical electrodynamics this result is natural, since gauge invariance forbids the existence of magnetic monopoles. But in 1931, Dirac showed that it is no longer true in quantum physics. There, the existence of monopoles is compatible with gauge invariance, but only if the strength of the magnetic monopole—the magnetic charge—has definite values. The product of the electric charge e and the magnetic charge g has to be a half-integer in natural units:

$$e \cdot g = \hbar c \cdot \tfrac{1}{2} n,$$

where n is an integer.

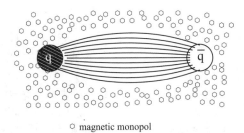

○ magnetic monopol

Figure 7.4. Compression of the color-electric field between a quark (q) and an antiquark ($\bar{q}$) by color-magnetic monopoles.

Not only does this equation imply that monopoles can exist, but it also indicates that if they exist, they solve an old puzzle: the existence of the universal elementary charge. If there exists one monopole with the magnetic charge g, then there must also exist an elementary charge $e = \frac{1}{2}\hbar c / g$ and all existing charges must be multiples of this elementary one.

In the grand unified theory, monopoles do not occur as elementary fields, but as solutions of the field equations. The field energy of these solutions, that is, the mass of the monopoles, is so large that they cannot be produced in accelerators, but possibly they occur in cosmic radiation. They could be leftovers of the big bang (see Section 7.6.2), and therefore monopoles were looked for in moon rocks. There are some events that can be explained by the occurrence of magnetic monopoles, but none of them have been confirmed beyond doubt.

Monopoles also play a decisive role in one of the most promising models for the permanent confinement of quarks in hadrons. S. Mandelstam and G. 't Hooft have independently proposed the following scenario: the ground state of QCD, the vacuum, consists of many color-magnetic monopoles. They cannot exist as free particles in the same way as quarks and gluons. They have the peculiar property of squeezing the color-electric fields together to form flux tubes. This is displayed graphically in Figure 7.4. This property has as a consequence that the color-electric force does not fall off with the inverse square of the distance, but rather the force has the same strength at all distances. Since work is given by force times distance, one needs an infinite amount of energy to separate a quark and an antiquark completely.

This effect has been observed in superconducting materials with the roles of electric and magnetic quantities interchanged. The ground state of a superconductor is formed of many electric charges in the lowest possible energy state (called Cooper pairs) that compress the magnetic field into flux tubes.

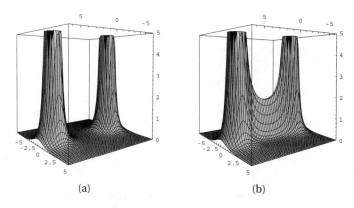

(a)   (b)

Figure 7.5. Model calculations of the field energy of a quark–antiquark pair. (a) Field energy as in QED, that is, without confinement. (b) Field energy in a model for QCD with confinement.

Numerical investigations of QCD on the lattice, as described in Section 6.7, support this picture of confinement, which sounds in this short and superficial description much vaguer than it is in reality.

One can extend this picture to more detailed models for quantitative calculations. Figure 7.5 shows the result of such a model for the density of the field energy of a quark–antiquark pair: on the left, in a field theory like QED, that is, without confinement; on the right, in a model for QCD, where a finite monopole density is assumed to form the ground state. The field energy is maximal near the quark and the antiquark. In the model for QCD, one sees a mountain ridge between the two quarks, which represents the energy density of the flux tube. The total field energy—the volume of the mountain ridge—increases with the distance between the quarks. A complete separation of the quarks (infinite distance) would require an infinite amount of energy.

7.6   The Microcosm and the Macrocosm

The same desire that drives us to extend the limits of our knowledge to the unimaginably small also drives our curiosity about the unimaginably large. It was Plato who introduced the universe (the All) as the fifth elementary particle, as shown in Figure 1.1. In the last decades a particularly close connection has arisen between particle physics and cosmology. On the one hand, the laws of particle physics are an important foundation for cosmological models, and on the other hand, cosmology sometimes offers the only possibility to check speculative assumptions in particle physics.

7.6.1 What We Do Know and What We Do Not (Yet) Know

The theory of general relativity is now the best-tested classical field theory. It is the firm basis of the standard model of cosmology. The triumphant advance of modern science began with the mathematical treatment of gravity in Isaac Newton's *Principia Mathematica Philosophiæ Naturalis* (1686). It was nearly another 200 years before electromagnetic phenomena could be explained as well as gravity. This was achieved in Maxwell's *Treatise on Electricity and Magnetism* (1876).

The complete mathematical formulation of electrodynamics in the framework of classical field theory had far-reaching consequences for our picture of the world. Max Planck concluded that the radiation laws of classical electrodynamics are incompatible with experience. This led him in 1900 to introduce the quantum hypothesis. It also turned out that the symmetries of the space-time transformations used in Newtonian mechanics are invalid in Maxwell's theory of electromagnetism. Therefore, Einstein developed the theory of special relativity, which was based on the transformation properties of electrodynamics. It revolutionized Newtonian mechanics. In the theory of general relativity, Einstein constructed a theory of gravitation that was compatible with the principles of special relativity but that went far beyond it.

Today, with the standard model of particle physics, we know the basic equations for the strong, electromagnetic, and weak interactions. Furthermore, at very small distances, there are indications that these interactions are different aspects of one and the same fundamental interaction. But we do not yet have a consistent quantized theory of gravitation, not even in the framework of perturbation theory. The main reason for that deficiency is presumably that we cannot perform dedicated experimental tests of such a theory. Quantum effects of gravitation are expected to be significant only at energies corresponding to the Planck mass, 10^{19} GeV. It is not exaggerated pessimism to assume that accelerators with this energy will continue to be unavailable even in the distant future.

However, in the distant past there was an event in which quantum effects of gravitation played a decisive role. It occurred at, and very shortly after, the big bang. We will discuss this event briefly in the next section.

7.6.2 Matter in the Universe

According to the present standard model of cosmology, our universe "came into being" 13.7 billion years ago through a kind of "big bang." This designation was coined by the astrophysicist Fred Hoyle as an invective, since he believed in a universe in equilibrium, in a steady state. There are two strong reasons for our belief that 13.7 billion years ago the universe was in an unusual state: first, we observe the detritus of the big bang in the background cosmic radiation, and second, the distribution of light elements can be explained only if the temperature of the universe, that is, the mean energy of its particles, was for some time extremely high. It is for that reason that in 1948, Alpher, Bethe, and Gamow predicted a big bang, long before the cosmic background radiation was detected by A. A. Penzias and R. W. Wilson in 1965.

I will not give a description of big bang cosmology and therefore will go only into the points closely connected with particle physics. To start on an optimistic note, we have good reason to believe that we know the natural laws governing what is by far the largest part of the history of our universe. When the universe had cooled off to a temperature of about 10 trillion degrees Celsius, the average energy of the particles was 200 GeV. This is a value that can be reached easily by present-day accelerators, and up to these energies, the standard model is certainly valid, at least to an excellent approximation. Thus from that time on, the development of the universe was governed by the laws of the standard model and those of the theory of general relativity. If one extrapolates the present state backward, this temperature was reached very shortly—about one-hundredth of a nanosecond—after the big bang.

At that time, the electroweak gauge symmetry was not yet broken. At these temperatures, the quadratic term with the "wrong sign" leading to the champagne-bottle potential of Figure 5.7(b) is more than compensated for by the contribution of the kinetic energy. But even after the Higgs mechanism became operative and particles acquired a mass, there were not yet hadrons. The strongly interacting quarks and gluons formed a plasma and were not yet confined inside hadrons. Hadrons could form only at temperatures corresponding to an average energy of 100 MeV, and such a temperature was reached approximately one-tenth of a second after the big bang. After about three seconds the universe had cooled off to a temperature of $\frac{1}{2}$ MeV, and particles such as electrons, protons, and neutrons "froze out." All unstable particles, such as the mesons, strange hyperons, and muons, had long since decayed.

The ratio of protons to neutrons of about six to one was determined by chemical equilibrium at the "freezing temperature" of about 1 MeV. From then on, during the first three minutes, the light primordial elements were formed. They range from heavy hydrogen (one proton and one neutron) to lithium (three protons and four neutrons).

After hydrogen, helium is by far the most abundant element in the universe. Nearly one-fourth of all nucleons in the universe are bound into helium nuclei. The reason for this abundance is the very high binding energy of the nucleons inside the nucleus. Lithium is very scarce: for each lithium nucleus there are more than two million hydrogen nuclei. From these numbers one can conclude that for each nucleon in the universe there are ten billion photons, but these photons contributed significantly to its (energy) density for only a very short fraction of the age of the universe. This is due to the photons' low energy. Because of the expansion of the universe, they have cooled off from a once dominant part of the universe to a meager background radiation, a photon gas with a temperature of only 2.7 degrees Celsius above absolute zero. This corresponds to an average photon energy of two-thousandths of an electron volt.

Conventional matter, of which we have certain knowledge, consisting of all the particles occurring in the standard model, contributes to only a small fraction of the density of the universe. Its complete density can be determined from different sources, for example, from the speed of the expansion of the universe and from the rotation of spiral nebulae. The conclusion from observations made in recent years is very exciting. Roughly, it can be summarized as follows: about 30% of the universe is matter, while the remaining 70% is *dark energy*. Matter makes itself felt through gravitation, dark energy through an acceleration of the expansion of the universe. The particles of the standard model constitute only 10% of this matter. The rest is unknown to us; it is *dark matter*.

All these numbers are taken from model calculations, but the fact that completely independent methods always lead to compatible numbers gives the model calculations a high degree of credibility.

In relativistic quantum field theory, one has no problem explaining the existence of dark energy; the problem is that there is too much of it. Quantum corrections to the vacuum energy as shown in Figure 1.18 are unavoidable, and they lead to an energy density of the vacuum. It is detectable only through its interaction with the gravitational field; therefore, it is considered dark energy.

In renormalized perturbation theory, there is no way to calculate these quantum contributions. In a field theory without gravity, the vacuum-energy density has no influence, and therefore it is impossible to fix it at a certain scale with a measured value. However, if one supposes that quantum field theory gives a realistic description of nature up to the smallest distances, then the calculated vacuum energy has to be taken seriously. If one assumes that the theory is valid down to distances corresponding to the Planck mass, then one obtains a value of the Planck mass to the fourth power for the energy density of the vacuum and, thus, the dark energy density Expressed in GeV per cubic femtometer, this value is 2.6×10^{59} GeV/fm^3. The observed value for the dark energy density is about 6.6×10^{-45} GeV/fm^3, that is, a discrepancy of 100 orders of magnitude, an absolute record in misjudgment.

Supersymmetry could in principle be the solution, but only in principle. In exact supersymmetry, the contributions of the bosons compensate for those of the fermions, and the resulting energy density is exactly zero. A small violation of supersymmetry could therefore lead to the small observed energy density. But unfortunately, supersymmetry is not weakly violated. From the fact that we have not yet seen supersymmetric partners of the known particles, one can conclude that the supersymmetry-breaking scale, E_{SUSY}, is at least 100 GeV. A simple calculation shows that the energy density of a broken supersymmetric theory is the breaking scale to the fourth power. With the value mentioned above, this yields an energy density of 1.3×10^{10} GeV/fm^3, still 55 orders of magnitude away from the observed value. Even if refined models can gain a few billion, one is still far away from the observed value. The problem of vacuum energy remains one of the biggest problems in physics, and it is certain that completely new ideas are needed to solve it.

In contrast to dark energy, there is perhaps already a solution for the problem of unknown dark matter. I mentioned in Section 7.4 that in the supersymmetric standard model there must exist a stable supersymmetric particle, presumably a neutralino, which interacts only very weakly with the known particles. This is an ideal candidate for dark matter. If it really exists, it should be detected in the next few years in high-energy experiments.

7.6.3 Recalcitrant Gravity

The theory of general relativity is a classical field theory of gravitation. It is, like electrodynamics, a gauge theory; its gauge group is the Lorentz group, that is, the transformations of space and time according to special relativity; the gauge field is the gravitational field.

A quantization of gravity in the framework of perturbation theory is formally possible. The field quantum of gravity must have spin 2. It is called the *graviton*. A perturbative treatment of gravity seems appropriate, since the coupling to gravity is very small in units appropriate to particle physics. Newton's gravitational constant has the value

$$G_N = 6.7 \cdot 10^{-39} \, \hbar c / (\text{GeV}/c^2)^2 = \hbar c / m_P^2.$$

In contrast to the gauge couplings of the electroweak and strong interactions, it is not dimensionless but has in natural units ($\hbar = c = 1$) the dimensions of an inverse mass squared. The mass m_P is the Planck mass, which we have encountered already on several occasions. The smallness of the coupling has the consequence that under normal conditions, quantum corrections to gravity are extremely small; they contain the factor $E_{\text{scale}}^2 \cdot G_N$, where the quantity E_{scale} is the energy (or the corresponding distance) that is relevant for the process under consideration.

Presently available energies are on the order of 1,000 GeV; this implies that quantum gravitational effects are suppressed even at these extremely high energies by the factor 10^{-32}, which is unmeasurably small. Shortly after the big bang, however, the energies were extremely high, and quantum corrections were important; therefore, a precise knowledge of quantum corrections is necessary if one is to make any well-founded statements about the big bang itself.

Such quantitative calculations cannot be performed, since a quantized perturbation theory of gravitation is not renormalizable. Even if one considers Newton's gravitational constant as a renormalized coupling, more and more infinities pop up that have to be eliminated by new conditions. This implies that gravity at extremely high energies, or correspondingly small distances, has to be modified. The situation is very similar to that of the weak interaction. Fermi's theory was as successful as Einstein's theory, but it was not renormalizable and had to be modified by the introduction of the intermediate bosons.

For some time, there was hope of solving the problem by supersymmetric gravity. In such a theory, the troublesome contributions of the gravitons are compensated for by those of their fermionic partners, which have spin $\frac{3}{2}$. It has not yet been proved that supersymmetric gravity is not renormalizable, but presently, two different approaches offer some hope. One consists in finding completely new approaches to quantization of gravity that are independent of perturbation theory. The other approach is what is known as string or superstring theory. It has been the focus of great interest in recent years, since it has yielded very important new mathematical insights.

7.7   Silent Strings

I cannot give here an introduction to, much less a review of, string theory. I will therefore simply mention some points related to particle physics.

String theory originated in the theory of strong interactions in the period of nuclear democracy. One of the main postulates of this approach was the *duality requirement*: the behavior of a scattering amplitude at large energies determines the behavior at large momentum transfer, and vice versa. In 1968, Gabriele Veneziano found a simple mathematical formula showing this behavior; In the Veneziano model, there were infinitely many hadrons, and the behavior at high energies was in accordance with the requirements of Regge theory (discussed in Section 3.4).

The Veneziano model was enthusiastically hailed in the community of particle physicists; one of them even spoke of the "Maxwell equations of particle physics." It was clear that Veneziano's formula could be only the first link in a chain. All hadrons in this model were stable, and—what was even worse— conservation of probability was violated. Attempts to correct these deficiencies showed that the model was mathematically closely related to the theory of vibrating strings. This was in accordance with ideas about mathematical quarks, that is, quarks that do not exist as free particles. The hadron was understood as a string, and the quarks corresponded to its endpoints. The end of a string cannot be isolated. If a string is cut, one obtains not two endpoints but two strings. In the same way, by "cutting" a hadron one obtains not quarks, but hadrons. In

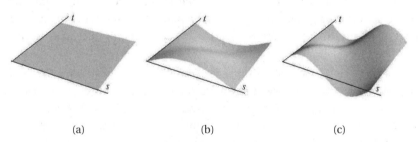

          (a)   (b)   (c)

Figure 7.6. (a) A nonexcited string (the ground state). (b) A simply excited string. (c) A doubly excited string. In the "old" string theory of strong interactions, the ground state was the lightest hadron of a series, and the vibrating strings corresponded to hadronic resonances. In modern string theory, the ground state has mass zero; the simply excited state, one Planck mass; and the doubly excited state, two Planck masses.

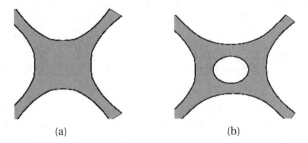

Figure 7.7. The interaction (scattering) of two strings. (a) The lowest approximation. (b) A contribution that corresponds to a quantum correction in quantum field theory.

this string picture of hadrons, the vibrational states of the strings correspond to hadronic resonances. The temporal development of vibrational states of a string is displayed in Figure 7.6.

The displacement of a string is a function of two parameters: time, in Figure 7.6 the t-axis, and the position along the string, the s-axis. The motion of a vibrating string is determined by the two-dimensional wave equation, a differential equation that was solved in 1747 by d'Alembert; it is still one of the few completely known general solutions of an important differential equation of mathematical physics. The explicit knowledge of the most general solution is essential for string quantization.

The string theory of hadrons describes the excitations (spectra) of hadrons and their interactions. The first approximation to the interaction of two hadrons is described by Veneziano's formula. In string theory it is represented in Figure 7.7(a). The two strings fuse to a single one and separate again. A correction to this first approximation is displayed in Figure 7.7(b). As with Feynman graphs, there are complicated mathematical expressions corresponding to these pictures.

The resulting mathematics is very interesting, but the physical results were rather disappointing. In the theory, there were states with imaginary mass, called *tachyons* (from the Greek tachys, swift), since they move with velocities faster than light in vacuum. Soon supersymmetry was applied to string theory, and string theory became superstring theory. Tachyons could be eliminated in superstring theory, but the theory contained massless particles of spin one and two. Their presence was again an unpleasant feature: the lightest hadron with spin 1 is by no means massless, and the observed mesons with spin 2, the tensor mesons, have an even larger mass.

There were some other features hardly compatible with the phenomena of hadronic physics. All this, together with the success of the field theory of strong interactions, QCD, led to a considerable downgrading of string theory in strong interactions. There are still attempts to explain nonperturbative effects with stringlike models (see Figure 7.4), but these attempts are much less ambitious than the hopes one had placed in the Veneziano model.

Yet the mathematical toy of superstring theory was much too interesting to abandon it completely. In 1974, J. Scherk and J. H. Schwarz made the theory more realistic by extending it beyond a theory only of hadrons, to a *theory of everything* (TOE). The massless particles were then highly desirable: those with spin 1 would be the gauge bosons of the standard model, while the massless particle with spin 2 would be the gauge boson of gravity, the graviton. This was—at least theoretically—the fulfillment of an old dream: a theory that described *all* interactions, including gravity. There were serious anomalies (see Section 5.8), but these could be removed by going from a world with three dimensions of space and one of time to a world with ten or twenty-six space-time dimensions. Of course, the dimensions beyond the four space-time dimensions of our everyday experience had to be compactified, as described at the end of Section 7.2.

The compactified dimensions have observable consequences. They are the degrees of freedom for internal symmetries such as the color SU(3). For some time, there was great optimism: many physicists believed that there were only a few symmetries that could be the consequence of a compactified string theory, and the standard model should be one of them. This hope has faded away. With explicit construction principles one could show that there is an astronomical number of possible symmetries and thus candidates for a "standard model."

The value of string tension in the old theory of hadrons was such that the excitations of the string were a few hundred MeV, the typical excitation energy of hadronic resonances. In modern string theory, the string tension is so high that the first excitation yields just the Planck mass. In the limit of an infinite string tension, the extended strings shrink to pointlike objects, and one obtains the only result of string theory that is comparable with experience: the resulting limit is a field theory with a local gauge symmetry. In the framework of superstring theory it is not an accident but a necessity that all fundamental interactions are gauge theories. But unfortunately, string theory cannot predict the concrete gauge groups. There are several other interesting results of string theory, but these concern the internal consistency of the theory rather than particle physics.

Jokingly, string theorists claim that their theory makes a large number of predictions that can—in principle—be checked experimentally. String theory predicts that there are not only the particles of the standard model, but that each of those has an infinite number of excited states. All fields of the standard model correspond to the strings in their ground state. In order to produce the predicted excited states, one needs an accelerator that attains energies of the Planck mass, that is, 10^{19} GeV. According to the present state of accelerator technique such an accelerator would require a circumference a billion times the length of the earth's orbit around the sun. Such an accelerator could not find enough space in our solar system. It was only for a very short time after the big bang that the superstrings were vibrating. If they were sounding then, they have long since fallen silent.

8

Epilog

In this epilog, I make some remarks on the position of elementary particle physics in relation to other branches of science and on the place of science in the human quest for knowledge. Up to now, I have tried to be as objective as possible and to represent nothing as certain that will not be regarded as certain in 100 years. This concluding chapter, however, will be subjective.

8.1 Peculiarities of Particle Physics

I have chosen a historical presentation for this book because I think that even the most recent developments can best be described in a (moderate) historical context. I also wanted to show that the development of particle physics was not a special path, but that it progressed in a rather straightforward manner from the given situation. Even developments that turned out to be dead ends have left their distinct traces. The assumptions made in the area of nuclear democracy, namely, that all hadrons are equivalent and that it makes no sense to look for more elementary particles, were finally replaced by the hierarchical field theory of QCD, in which the fields of quarks and gluons are privileged. Nonetheless, the observable hadrons are still equivalent, and the quarks are not constituents of the hadrons in such a direct sense as the electrons and nucleons are constituents of atoms.

Particle physics is unique in that its experiments are on a scale that is larger than in almost every other branch of science. The devices in use are truly gigantic, even compared with large-scale devices of the first half of the twentieth century. This was mentioned in the introduction and shown in Figure 1.2. That this is the case is due not to an outbreak of particle physicist megalomania, but to the nature of the field. As explained in Section 5.7, the resolution of ex-tremely small structures needs high energies, and thus huge accelerators and measuring instruments.

The size of the experiments has two consequences: first, the costs are very high, and second, the way that science is done has changed in a fundamental way. The number of people collaborating in an experiment is very large. Recall the earlier mention that the list of authors in the papers on the detection of the top quark took nearly as much space as the whole article of Neddermeyer and Anderson on the discovery of antimatter.

Both points are intensely discussed by sociologists and science administra-tors (with apparently different emphasis on the two points). I want to make a few remarks from the point of view of a participant, even if as a theoreti-cian I am not directly concerned with these problems. Experiments and costs are large, but the community of particle physicists has reacted very reasonably. Nearly all projects, even those at national laboratories, are carried out as inter-national collaborations. This has very good consequences for the education of young scientists, and it also leads to a quite reasonable international coordina-tion of necessary experiments. At CERN, it has been definitively demonstrated that even a European organization can work effectively. This has nothing to do with particle physics, but can be regarded as a beneficial political side effect. The large number of persons collaborating in one experiment also keeps the expenditure per person reasonable. The costs for a dissertation in high-energy physics are therefore comparable to those in any other field of the experimental sciences.

At issue is not whether the expenditures for particle physics can be justified by some direct practical benefit. Rather, it is a question of culture: what finan-cial resources does a society make available for fundamental research? I have pointed out that the quest for knowledge concerning the smallest constituents of matter has a continuous history in Western culture. One should also be aware that the practical use of new knowledge is always difficult to imagine. Reportedly, when asked by the British prime minister Disraeli about the prac-tical use to which electricity would be put, Faraday replied, "I do not know, but I am sure that one day government will put taxes on it."

The indirect profits in the fields of information technology are more evi-dent. As an example, I mention only the World Wide Web, which was con-ceived and developed at CERN to facilitate communication within the large group of collaborators—often dispersed across the globe—participating in an experiment.

As a university teacher I also have seen that a degree in particle physics has value outside the immediate field. I will leave a more detailed discussion of these important questions to the specialists at the large research centers.

Concerning the large number of physicists who collaborate on a single experiment, it seems that a reasonable division of labor is possible. Within a large experiment, a small group or even a single person is responsible for a well-defined task. Examples of this are the analysis of a certain production process or the design and construction of a Cherenkov threshold counter. Since the result of a large experiment depends on all its components, it is reasonable that all those involved appear as authors of the article announcing the result, such as the detection of the top quark. It is, of course, a difficult task to maintain a satisfactory overview of a complex experiment from design through data analysis. This explains, for example, why the awarding of the Nobel Prize to Rubbia and van der Meer was not criticized within the large group of participants.

I will not discuss the sociology of particle physicists, except to mention that investigations in this area have generally concluded that scientists working in this field are normal people with the more or less amiable weaknesses displayed generally by the species *Homo sapiens.*

An important issue in science is that of the reliability of experiments. Here the problem seems to be that the large size and resulting cost of an experiment often makes it very difficult to check the results of such an experiment directly. Nevertheless, I know of no cases of forgery, and gross errors are very rare. This is a result of strong control inside the collaboration, whereby a large number of specialists working in related fields are eager to find any fly that may have wandered into the ointment, particularly in the form of an error in the result of a different group. Colleagues from experimental physics report that the critiquing process within the collaboration is normally much more severe than in the refereeing process. Internal criticism is further promoted by the fact that in a huge experiment, different groups work on the same problem with different methods, such as different types of counters. An incorrect experimental result is a black mark for an experimentalist, much more than a theory later shown to be wrong is for a theoretician. This is a good thing, since Bacon's remark, quoted earlier, that truth arises rather from untruth than from confusion is true only for theoretical speculations and not for solid measurements.

Experimental error is nonetheless never totally excluded. Serious errors—those with far-reaching consequences—are normally corrected quickly. An example of this process is the treatment of experimental results from the electron accelerator CEA, in Cambridge, Massachusetts, that contradicted QED. The re-

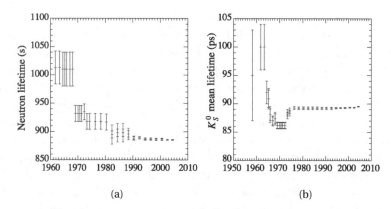

Figure 8.1. Published results in *Review of Particle Physics* (a) for the lifetime of the neutron and (b) for the short-lived component of the K^0 meson over time.

sults were published in 1965 in *Physical Review Letters* under the title "Deviation from Simple Quantum Electrodynamics." The measurements were repeated at DESY by S. C. C. Ting and his collaborators and led in 1968 to a publication in the same journal, but this time with the title "Validity of Quantum Electrodynamics at Extremely Small Distances." Theory had won the day. Erroneous results having little influence might survive a bit longer.

Figure 8.1 shows two results that have changed over the course of time by more than (simple) statistical errors. Figure 8.1(a) shows the results for the lifetime of the neutron; Figure 8.1(b), the results for the lifetime of the K^0 meson. Evidently, a systematic error had been in effect before 1966 for the neutron, and for the K^0 meson, the systematic errors had been underestimated. Since 1993 there has existed a discrepancy of more than 10% between two results for the cross section of proton–antiproton scattering at very high energies. Most probably this is not a statistical fluctuation but due to a yet unknown systematic error.

Since the beginning of empirical science it has been clear that one has to look for *disturbance effects* in an experiment and correct for them. A famous example is the correction for air resistance in freefall. The same holds for particle physics. The disturbance effects and the peculiarities of the measuring device are here often taken into account in a very special way, namely, through *Monte Carlo simulation*. The name derives from the stochastic character of this method: one assigns the events a certain probability and then, so to speak, plays roulette with them by calculating the events with the given probability distribution many times, thereby obtaining the expected distribution.

There are two reasons for using such statistical methods. First, it is often difficult or even impossible to calculate a process fully deterministically, even for processes to which the laws of classical physics can be applied safely. For example, if one wants to determine the outcome of a throw of a die by the laws of classical mechanics, this might be possible in principle, but it is very difficult. If one is not interested in the result of a single throw, but only in the distribution of many, then a simple argument taking into account the symmetry of the die allows one to predict that the probability of throwing any given number between 1 and 6 is 1/6.

Second, and this is even more important, the disturbance effects are mostly determined by quantum mechanics, and a classical deterministic description is impossible. Since quantum mechanics makes predictions on probabilities, Monte Carlo methods are necessary. It is possible to describe quantum-mechanical events by classical statistics if one gives up the requirement of locality.

Monte Carlo simulations are used for more than calculating disturbance effects and correcting for them. They are important for the analysis of the data

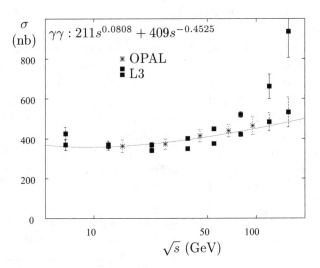

Figure 8.2. The cross section σ for scattering of photons on photons as a function of the energy $\sqrt{s}$ as obtained from two different groups. The differing results from the group L3 (solid squares) were obtained using different Monte Carlo programs.

itself. In many experiments, only a small fraction of the interesting results can be detected in the counters, but one is interested in the total number of such events. A notorious example is the scattering of photons on photons, which can be investigated in electron–positron scattering. Here the total number is reconstructed from the measured one with the help of a Monte Carlo program.

Unfortunately, the results of a measurement thus obtained depend strongly on the Monte Carlo program used. An example is displayed in Figure 8.2. The "experimental points" with error bars come from two different groups, called L3 and OPAL, both performing their experiments at the large electron–positron collider LEP at CERN. The results of the group L3, the solid squares in Figure 8.2, were obtained with two different programs, leading at high energies to drastically different results. The errors displayed in the figure are only the statistical ones; the final errors also have to include systematic uncertainties such as those due to the Monte Carlo programs. The photon–photon scattering is admittedly an extreme case, since the programs used were not written especially for this reaction. Nevertheless, it shows how strongly experimental results can be influenced by theoretical preconditions. If the curve included in Figure 8.2 were the result of an otherwise well-tested theory, one would claim excellent agreement between theory and experiment and no longer use the program that leads to the values that are too high.

Monte Carlo programs also play an important role in comparing experimental results with theoretical calculations for hadronic processes. I mentioned in Section 6.6 that certain calculations can only be performed for quark reactions, and not for hadronic ones. Here, too, one uses Monte Carlo methods in order to obtain cross sections for hadrons from those for quarks.

Particle physics deals with phenomena that are very distant from our daily experience, and this has consequences for its relationship to other branches of physics. It is scarcely conceivable that results of particle physics will one day have some sort of direct influence on, say, chemistry or biology. The only direct connection now is with cosmology and this connection is so tight that both are nearly fused into a single field, as outlined in Section 7.6.

Another close connection emerged in condensed-matter physics. However, this refers not to the objects of investigation but to theoretical methods. Concepts like spontaneous symmetry breaking and the renormalization group play a central role in both fields and were developed in a close interplay between the fields, sometimes by the same individuals.

This tight connection seems surprising at first. It has its origin in a special approach to quantum field theory developed by Feynman. It has become now, especially in gauge theories, the standard method. I cannot deal with this in detail and will content myself with the somewhat vague remark that quantum field theory is formally the theory of condensed matter in four space dimensions. For that, one has to choose the time in field theory as imaginary. That is, one has to replace t by $i\,t$ in the formulas. The calculations on the lattice as described in Section 6.7 are all performed in such a four-dimensional space with imaginary time. In condensed matter physics, the role of imaginary time in field theory is played by the inverse temperature, $1/T$.

Intuitively, one can use the following superficial picture: in quantum corrections, the production of virtual particles is key, and each problem of quantum field theory is therefore a problem in which many (virtual) particles occur. The shorter the time, the more virtual particles can be produced; the energy–time uncertainty makes the restrictions of energy conservation less and less stringent. In condensed matter physics, there is a correspondence between this production of virtual particles and the excitation of states with higher energy. At high temperatures—small inverse temperature—many highly energetic states can be excited. High temperature in condensed matter physics corresponds to small time intervals, and therefore we have formally a similar situation. The excited states in condensed matter are real, whereas the produced particles in quantum field theory are virtual; this is a consequence of imaginary time. I want to emphasize that well-defined mathematical concepts and operations corroborate these vague words.

8.2 Philosophy

The question of whether we can gain philosophical insight from particle physics is difficult to answer. If there is an answer, it will be subjective. The natural sciences have fared well in recent centuries by emancipating themselves from philosophy. In S. Weinberg's book *Dreams of a Final Theory*, there is a chapter entitled "Against Philosophy." However, it was an eminent scientist, H. Helmholtz, who stated, "it will always remain the business of philosophy to investigate the sources of our knowledge and the degree of its justification; no century can evade this assignment and go unpunished."

In fact, every scientist has his or her own private philosophy, which the scientist follows more or less, mostly less. Einstein openly advocated an epistemological opportunism, and the philosopher Ernst Cassirer remarked that the scientific actions of many physicists show a common inner continuity and consistency, even though their (philosophical) judgments about these actions vary widely.

In my opinion, the soundest epistemological basis for modern science is that of a symbolic knowledge of nature. This program was very clearly formulated by Heinrich Hertz, the discoverer of electromagnetic waves: "We form for ourselves inner phantoms or symbols of things in such a way that the logical consequences of the symbols are always pictures of the physically necessary consequences of the depicted objects." This was elaborated by the philosopher Ernst Cassirer and strongly advocated by Hermann Weyl, the master of the gauge (see Section 5.1). I would like to demonstrate briefly the utility of this approach to some aspects of particle physics.

At present, the following principles are the basis of particle physics: the concepts of quantum physics, the locality postulate, and the gauge principle. Locality and gauge invariance originated in special and general relativity; quantum physics, in atomic physics and statistical mechanics. If we compare the scales that dominated the origins of these concepts with those that are resolved today in particle physics, then the difference in scale is nearly as big as that between atoms and macroscopic bodies. The step from the t quark to the atom is not quite as big as that taken by Newton, who drew conclusions about celestial bodies from observing falling apples, but it is comparable.

Our scientific modes of explanation have not yet reached their limit. We have good reason to believe that we know not only the general principles, but also the concrete microscopic laws, that govern the interaction and thus the structure of matter known to us. These interactions are the strong and weak interactions, which determine the structure of atomic nuclei and thereby the existence of chemical elements, and the electromagnetic interaction, which is responsible, among other things, for all chemical reactions. In all these cases, the use of quantum field theory is imperative for an adequate description. Only through quantum field theory can such elementary processes as emission or absorption of light by atoms be described consistently. Furthermore, the Pauli exclusion principle, at first an empirical rule, turned out to be a consequence of local quantum field theory.

The structures of these three interactions show remarkable similarities; all three obey the gauge principle. It is possible that in a few years we will know with certainty that all three interactions can be reduced to one. According to our present knowledge, a supersymmetric unified gauge theory is the hottest candidate for such a theory. The gauge principle goes beyond the strict principle of locality, since it includes the connection of matter fields at neighboring points through the gauge fields. In a gauge theory on the lattice, as discussed briefly in Section 6.7, this can be seen from the fact that the gauge fields are assigned to the links between the points, and not the points themselves.

It might be possible that with higher resolution one will find that it is not local fields but extended strings that are the fundamental objects of an adequate description of microscopic physics. We would then have to invent and use a new set of symbols without invalidating the results of the standard model.

In the early 1990s, there was hope that all the riddles of particle physics would soon be solved (in principle), and Weinberg dreamed his "Dream of a Final Theory." Today, one's hopes are more modest. An old concept was rediscovered and formalized, the *effective theory*. In an effective theory, the limits are built in. A very good example is Fermi's theory of beta decay. In the standard model, the decay of, say, a muon into an electron, a mu neutrino, and an anti-e neutrino proceeds through the exchange of a W boson; see Figure 8.3(a). The spatial resolution is determined by the inverse of the momentum transferred from the muon to the electron and its anti-e neutrino. The muon cannot transfer more momentum than it has rest energy, and therefore the spatial resolution is proportional to the inverse mass of the muon. The exchanged W boson is 762 times more massive than the muon, and therefore the range of the interaction is smaller than the spatial resolution of muon decay by this factor. Therefore, to very good approximation one can consider the interaction to be pointlike.

This is just what Fermi assumed in his four-fermion interaction (Figure 8.3(b)). Fermi's theory is an effective theory for weak decay in which the trans-

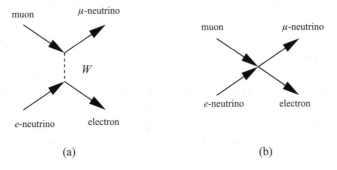

(a) (b)

Figure 8.3. (a) Graph for the decay of a muon in the standard model. (b) Graph for the decay of a muon in Fermi's effective theory. The incoming e neutrino line describes an outgoing anti-e neutrino.

ferred momentum is small—in natural units—compared to the mass of the intermediate boson. The error incurred by neglecting the detailed structure of the interaction is approximately proportional to the square of the mass ratios, in the above-mentioned case about two-thousandths of one percent. However, if one performs an experiment in which the momentum transfer is several GeV, or if one can measure the experimental values extremely precisely, the effective theory becomes insufficient.

Historically, the course of events followed precisely the reverse order. There was first a very successful theory with some theoretical flaws, and later it was found that it was an effective theory limit of a renormalizable theory. It might well be that the standard model is an effective theory for a supersymmetric grand unified theory or that field theory in its entirety is an effective theory of string theory.

Without disparaging the large formal apparatus of effective theory, one may say that an effective theory is one that makes reliable predictions in a limited range, or to phrase it differently, we have found a picture that meets the requirements proposed by Hertz mentioned above.

A special problem in particle physics is the gap between the objects seen in experiments, the particles, and the concepts used in theory, the quantum fields. I do not want to expand upon the problem and go back to the famous "wave–particle dualism" or discuss the serious problems of the particle concept in a possibly quantized theory of general relativity; I will start immediately with quantum physics. Even if we proceed very pragmatically, we are confronted with several problems.

We begin with a definition for particles that is close to experiment. It goes back to E. Wigner. According to this definition, a particle is a state with a fixed discrete mass value that can be determined by the energy and the momentum. A zero value for the mass is admissible, and as discussed in Section 7.2, this is actually the most natural value for the mass of an elementary particle. The particles that have been known for the longest time, such as the electron, proton, and photon, are adequately subsumed in this definition. For the neutron or even the pi meson, the lifetime is so long that we can speak in good conscience of a well-defined mass value. This holds even for the neutral pi meson, whose lifetime, about 10^{-16} seconds, leads to a mass uncertainty of about a hundred-thousandth of one percent. The transition is fluid: the J/ψ-meson has a width of about 0.004% of its mass, the phi meson 0.4%, the omega meson about 1%, and the rho meson 20%. Generally, all these states are called "particles"—at least they are listed as such in the often cited *Review of Particle Physics*.

Fortunately, the question of what a particle actually *is* is irrelevant in particle physics. What is important is the choice of adequate symbols to describe an experimental situation or to build a theory. If in the course of planning or carrying out an experiment one speaks of particles, then one means states that live long enough that one can construct beams of them or detect their traces. This implies that the lifetime is longer than about one picosecond (10^{-12} seconds). In the course of analysis, one is more liberal. One speaks of the production of a rho meson if one detects the traces of its decay products, the pi mesons, in the correct energy range.

In theoretical descriptions, quantum fields are the adequate quantities. The relation, however, between fields and particles is not so simple, for it depends on the chosen picture. If one confines oneself to free fields, the relation is well defined. One constructs a space of states for the particles, and the fields are constructed of the operators that create and annihilate the particles in that space.

This relation remains valid in perturbation theory. Thus in perturbation theory, the relation between the photon and the electromagnetic field is well defined, as is the relationship between the gauge fields and the intermediate W and Z bosons in weak interactions, though there the width is already three percent of the mass. However, one cannot see the W and Z bosons directly, such as by their traces, since the width implies a lifetime of about 10^{-24} seconds.

Theoretically, however, it is not important how long a particle lives, only whether it is meaningful to use perturbation theory. This is the case for the electromagnetic and weak interactions. However, not only is the assertion "there exists in the framework of quantum field theory an elementary field whose field quanta cause predictable effects" more precise than the statement "W bosons exist," it also has more consequences. The quantum field has effects not only in the final state, when the W boson is produced as a free particle, but also in the quantum corrections, where it makes itself felt in the field propagator.

The situation is more complex for the hadrons. The proton is an impeccable particle according to Wigner's definition, but there is no corresponding fundamental quantum field. Since 1975, we have known—or phrased more cautiously, we have excellent reasons to believe—that hadrons are composed of quarks. In somewhat more precise language, this means that fields for hadrons can be constructed from fundamental quark fields. We have several reasons for making this statement. Under certain conditions, we can draw observable consequences from it with the help of perturbation theory, for example in deep inelastic scattering. But since gluon and ghost fields also occur in relativistically invariant perturbation theory, we have to ascribe to them as much—or as little—reality as we do to quarks. Fortunately, we need not rely completely on perturbation theory; for instance, we can calculate masses and other properties of hadrons in lattice QCD, at least numerically.

We are now prepared to tackle the question, "Do quarks exist?" The difficulty here lies not in the concept of quarks, but in the meaning of existence. It is certainly true that quarks as isolated particles have not been found, in contrast, say, to electrons and protons. If we use any of the above-mentioned definitions of a particle, then we must say that quarks have not been detected as particles. This is not to be taken as a poke in the eye of the standard model. On the contrary, one does not expect the occurrence of free quarks. Quark *fields*, however, play an important role in the standard model. If we consider their occurrence in a consistent theory that has many observed consequences as proof of existence, than we can say that indeed, quark fields exist. However, this statement relies heavily on a sophisticated theory, and therefore Gell-Mann's description of quarks as *mathematical objects* is the most appropriate. In the above-mentioned symbolic representation of nature, the question is easily answered: quarks are essential symbols in the description of subnuclear phenomena. They are thus indispensable—at least at the present stage of our knowledge.

A

Glossary

References refer to places in the book where an expression is introduced or discussed in detail.

abelian group. In an abelian group of transformations, the result of two successive transformations does not depend on the order. Rotations in the plane and translations form an abelian group, while rotations in three-dimensional space are nonabelian ((after the mathematician N. H. Abel). ....Section 1.5.1.

alpha particle. The atomic nucleus of helium. It consists of two protons and two neutrons. In radioactive alpha decay, the decaying nucleus spontaneously emits an alpha particle. ....Section 1.3.

annihilation operator. An operator that annihilates a state. A quantum field is composed of creation and annihilation operators. ...... Section 1.4.2.

anomaly. A quantum correction that violates a symmetry (from Greek *anomos*, against the law). ....Section 5.5.

antimatter. Matter consisting of antiparticles. .... Section 1.4.2.

antiparticle. According to relativistic quantum field theory, for each particle there exists a corresponding antiparticle with the same mass and spin, but opposite charge and baryon or lepton number. Antiparticles are indicated by the prefix anti-, e.g., antiproton, anti-K meson. The antiparticle of the electron is the positron. ....Section 1.4.2.

atom. Since antiquity, the name given to the ultimate indivisible constituents of matter. According to modern usage, the atoms are the building blocks of the chemical elements. Atoms consist of an atomic nucleus and an atomic shell (or cloud) of electrons. The *nucleus*, in which nearly all the mass of the atom is concentrated, consists of protons and neutrons. In the case of hydrogen, there is only a proton (from Greek *atomos*, indivisible). . . . . Section 1.2.

axial vector. A vector that, under rotations, behaves like a "normal" oriented quantity (vector), but that does not change its sign under space reflections. It is also called a *pseudovector*. . . . . Section 1.5.4.

baryon number. A property of elementary particles introduced in order to explain the stability of protons. Particles with baryon number $B = +1$ are called *baryons*; those with $B = -1$, *antibaryons*. Baryons are strongly interacting particles (from Greek *barys*, heavy). . . . . Section 1.5.4.

beta decay. Radioactive decay of an atomic nucleus through the emission of an electron and an antineutrino. . . . . Section 1.2.

BNL. Brookhaven National Laboratory, an accelerator center near Brookhaven, in Long Island, New York. . . . . Section 2.6.

Bose–Einstein statistics. A state obeys Bose–Einstein statistics if it remains unchanged under the exchange of two identical components. States obeying these statistics are not subject to the Pauli exclusion principle (after S. N. Bose and A. Einstein). . . . . Section 2.4.

boson. Particle with integer spin. Bosons obey Bose–Einstein statistics and are not subject to the Pauli exclusion principle. . . . . Section 2.4.

CERN. European accelerator center near Geneva, Switzerland (from French *Centre Européen de la Recherche Nucléaire*, European Organization for Nuclear Research). . . . . Section 2.6.

charge conjugation. A transformation of particles into antiparticles. Neutral states have an internal *charge parity* of plus or minus one under charge cojugation. . . . . Section 1.5.4.

charm. A chargelike property of quarks and hadrons that is conserved in strong and electromagnetic interactions. *See also* **flavor.** . . Section 6.3.1.

chirality. Handedness. A particle is called right-handed if spin and motion are parallel; otherwise, it is left-handed. . . . . Section 2.8.1.

chiral symmetry. If this symmetry is valid, then right- and left-handed objects transform independently. Only massless particles may have this symmetry (from Greek *cheir*, hand). ....Sections 2.8.1, 7.2.

classical physics. The physics that was assumed to be generally valid before about 1900 and in which quantum effects are not taken into account. ....Section 1.1.

color. Chargelike property of all quarks. *Color SU*(3) is the gauge symmetry of color charge. ....Section 4.4.

commutation relation. Algebraic relation between operators **A** and **B**, the result of the *commutator* $[\mathbf{A}, \mathbf{B}] = \mathbf{AB} - \mathbf{BA}$. ....Section 1.4.2.

complex number. A generalization of real numbers, constructed from real numbers with the addition of the imaginary unit i, defined to be the square root of -1; that is, $i \cdot i = -1$. Every complex number is determined by two real numbers a and b, and is represented as $a + b \cdot i$. Another important representation is in terms of the positive modulus and an angle between 0 and 360 degrees. The relation between the two representations in the *complex plane* is shown in Figure A.1.

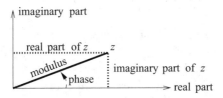

Figure A.1. Representation of a complex number in the complex plane.

Under *complex conjugation*, marked by an asterisk *, the imaginary part changes its sign: $(a + ib)^* = a - ib$. ....Section 1.5.2.

conservation law. If a quantity such as energy is conserved, it remains unchanged over time and during reactions. Some quantities are conserved only by certain interactions, and broken or violated by others. Parity, for example, is conserved only by strong and electromagnetic interactions, but not by weak interactions. ....Section 1.5.1.

cosmic rays. Components of radiation from outer space. They consist partly of highly energetic particles, mainly protons and alpha particles (known as the primary component). These make secondary particles in the atmosphere such as pi mesons. .... Section 1.6.

creation operator. Operator that creates a state. A quantum field is composed of creation and annihilation operators. .... Section 1.4.2.

cross section. Number of scattered particles relative to one scattering center and a unit current of incoming particles. It is the surface of the target that a scattering particle "sees." .... Section 1.2.

cyclotron. Circular particle accelerator (from Greek *kyklos*, circle). . . Section 1.7.

DESY. German accelerator center in Hamburg, Germany (From German *Deutsches Elektronen Synchtrotron*, German Electron Synchrotron). Section 1.1.

Dirac equation. Relativistic quantum wave equation for spin-$\frac{1}{2}$ particles (after P. A. M. Dirac). .... Section 1.4.1.

electron. Negatively charged particle; it is a part of the atomic shell. The electron has lepton number 1 and baryon number 0 (from Greek *elektron*, amber). .... Section 1.2, Table 6.2.

electron volt. Unit of energy, abbreviated by eV. Its multiples are MeV = $1,000,000$ eV and GeV = $1,000$ MeV. One electron volt is the kinetic energy gained by an electron—or any particle with elementary charge 1—if it passes through a voltage difference of 1 volt. .... Section 1.1, Table B.1.

elementary charge. Electric charge of the proton. The electron has elementary charge -1. .... Section 1.1, Table B.1.

family. Two quarks of different flavors and two leptons of different charges are grouped into each family. One example of a family includes the electron, the *e* neutrino, and the *u* and *d* quarks. .... Section 6.3.1, Table 6.2.

Fermi–Dirac statistics. A state obeys Fermi–Dirac statistics if it changes sign under the exchange of two identical components. States with these statistics are subject to the Pauli exclusion principle (after E. Fermi and P. A. M. Dirac). .... Section 2.4.

Fermilab. Accelerator center near Chicago, Illinois (after E. Fermi). .... Section 6.8.

fermion. Particles with half-integer spin. Fermions obey Fermi–Dirac statistics and are subject to the Pauli exclusion principle. ....Sections 1.2, 2.4.

Feynman graph. Graphical representation of probability amplitudes for reactions in quantum field theory (after R. Feynman). .... Section 1.4.2.

field. Assignment of physical properties to space-time points. .. Section 1.4.1.

field quantum. A particle assigned to a quantum field that consists of the creation and annihilation operators of just this particle. The photon, for example, is the field quantum of the quantized electromagnetic field. ....Section 1.4.2.

flavor. Quantum number distinguishing the six quarks. The flavors include strangeness, charm, beauty, and topness. *Flavor* $SU(3)$ is the broken symmetry due to the nearly equal masses of the u, d, and s quarks. .... Section 4.2.

gauge symmetry. Class of symmetry in which the symmetry transformations can be different at each point in space and time. *Gauge fields* are the fields that ensure *gauge invariance*, that is, invariance under these symmetries. The *gauge bosons* are the field quanta of the gauge fields. .... Sections 5.1, 5.2.

generator. From the generators of a group all group elements can be generated. ....Section 1.5.1.

gluon. The field quantum of the gauge fields of quantum chromodynamics. It has spin 1 and is, like the quarks, confined inside hadrons (from *glue*). ....Section 6.4, Table 6.3.

Goldstone boson. Massless boson appearing upon spontaneous breaking of a symmetry (after J. Goldstone). .... Section 5.3.

group. Set of mathematical objects with certain properties. ..... Section 1.5.1.

hadron. Name for strongly interacting particles (after Greek *hadros*, hard). ....Section 3.3, Table 6.5.

Higgs boson. Boson responsible for generating the masses of the gauge bosons in the sector of weak interactions (after P. W. Higgs). .. Section 5.4.

hypercharge. Denoted by Y. For strongly interacting particles the sum of baryon number and strangeness; left-handed leptons have *weak hypercharge* $Y = -1$; right–handed leptons, $Y = -2$ (from Greek *hyper*, above). .... Section 2.2.

imaginary number. *See* **complex number.**

interaction. In particle physics we deal with three interactions: the *electromagnetic interaction*, which we know from everyday experience; the *weak interaction*, which is responsible for radioactive beta decay; and the *strong interaction*, which acts within the nucleus. ..... Section 1.4.2.

intermediate bosons. Gauge bosons of the weak interaction; they have spin 1. The charged intermediate bosons are called $W^{\pm}$ bosons; the neutral one is called the Z boson................................ Section 6.1, Table 6.3.

ion. Electrically charged atom or molecule (from Greek *ion*, something that goes). .... Section 1.3.

isospin. Internal symmetry of elementary particles. .... Section 1.5.3.

Lagrangian. Expression from which the equations of motion in mechanics and field theory can be derived. It is closely related to the energy density (from J.-L. Lagrange). .... Section 6.2.

lepton. Name for only weakly interacting particles with spin $\frac{1}{2}$ (from Greek *leptos*, tender). .... Section 2.1, Table 6.2.

locality. Postulate (and its consequences) that no interaction can propagate faster than light in vacuum. .... Section 1.4.1.

magnetic moment. Many particles are magnetic dipoles and therefore have a magnetic moment. .... Section 1.2.

matrix. Mathematical object consisting of complex numbers, for example, arranged in a rectangular scheme. There are fixed rules for the algebra of matrices (from Late Latin *mater*, breeding animal). .... Section 1.5.1.

meson. Hadron with baryon number zero. Originally the muon, a lepton, was called the mu meson (from Greek *mesos*, middle). .... Section 1.4.2.

mesotron. Original name for particles heavier than electrons and lighter than nucleons. .... Section 1.5.3.

momentum. Oriented quantity characterizing the state of motion of a particle. The direction of momentum is given by the direction of motion. The modulus of the momentum of a particle with mass m and velocity v is given by $p = m \cdot v / \sqrt{1 - v^2/c^2}$; the relation between the total energy E (including the rest energy $m \cdot c^2$) and the momentum is given by $E = \sqrt{m^2 \cdot c^4 + c^2 \cdot p^2}$. The total momentum of a system is a conserved quantity. In the nonrelativistic limit v/c is small and $p \approx m \cdot v$ and $E \approx mc^2 + \frac{m}{2} v^2$. .... Section 1.4.1.

muon. Charged lepton similar to the electron, but heavier (from the Greek letter μ, *mu*). .... Section 2.1, Table 6.2.

neutral currents. Terms in the weak interaction that do not change the charge of the interacting fermions; they are due to the exchange of virtual intermediate Z^0 bosons. .... Sections 5.4, 6.3.

neutrino. Neutral lepton that appears together with the charged leptons in weak interactions (from neutron with the Italian suffix of diminution *-ino*, small) .... Section 14, Table 6.2.

neutron. Electrically neutral particle. Together with the proton, a constituent of the atomic nucleus (from Latin *neuter*, neither). .... Section 1.2.

nonabelian group. A group of transformations in which the result of two successive transformations depends on the order in which the transformations are made. *See also* **abelian group.** .... Section 1.5.1.

nucleon. Common name for the constituents of the atomic nucleus, namely the proton and the neutron; generalized to all baryons with strangeness 0 and isospin $\frac{1}{2}$ (from Latin *nucleus*, kernel). .... Section 1.5.3.

nucleus *See* **atom.**

parity. Determines the behavior of a state under space reflections, also called *parity transformations*. The parity of a state is $+1$ if it remains unchanged under space reflections, −1 if it changes sign. The *internal parity* of a particle is determined by the behavior of its quantum field. The total parity of a state is the product of the external and internal parities (from Latin *paritas*, equality). .... Section 1.5.4.

parton. Constituent of hadrons in a model developed for scattering by R. Feynman (from Latin *pars*, part). .... Section 5.8.

Pauli exclusion principle. This principle states that in a state with two fermions, the fermions must have different quantum numbers. It is a consequence of the theorem on spin and statistics. ..... Sections 1.2, 2.4.

phase factor. Complex number of modulus 1. .... Section 5.1.

photon. Field quantum of the electromagnetic field, also called the *gamma quantum* or light quantum; it is the gauge boson of electromagnetic interactions (from Greek *phōs*, light). .... Section 1.2, Table 6.3.

Planck constant. Elementary quantum of action, a fundamental natural constant that always occurs if quantum effects are important; its symbol is h. The *reduced Planck constant* is $\hbar = h/(2\pi)$. .... Section 1.1.

positron. The antiparticle to the electron. .... Section 1.3.

probability amplitude. A complex number, the absolute square of which is the probability of a reaction. .... Sections 1.4.2, 3.2.

proton. Positively charged particle; together with the neutron a constituent of the atomic nucleus (from Greek *prōton*, the first). .... Section 1.2.

pseudovector. *See* **axial vector.**

QCD. Abbreviation for quantum chromodynamics.

QED. Abbreviation for quantum electrodynamics.

quantum chromodynamics. The quantum field theory of colored quarks and gluons, abbreviated QCD (from Greek *chrōma*, color, and Greek *dynamis*, force). .... Section 6.4.

quantum corrections. Typical effects of quantization of a field theory, caused by virtual particle creation and annihilation. .... Sections 1.4, 2.4.

quantum electrodynamics. Quantum theory of electrodynamics, abbreviated QED. . . . . Sections 1.4.2, 2.4.

quantum number. Properties of states, especially particles, that take discrete values; examples are spin and internal parity. . . . . Section 1.5.4.

quark. Elementary constituent of hadrons. Quarks have spin $\frac{1}{2}$, interact with gluons, and are confined. . . . . Section 4.3, Table 6.2.

radioactive decay. Decay experienced by some naturally occurring and practically all artificially produced atomic nuclei. One distinguishes three kinds of radioactive decay, according to the kind of particle emitted by the mother nucleus: 1. alpha decay: a helium atom, also called an alpha particle; 2. beta decay: an electron and an antineutrino or a positron and a neutrino; 3. gamma decay: a photon, also called a gamma quantum. . . . . Section 1.2.

relativity, theory of special. Theory developed by Einstein in which space and time coordinates are transformed in such a way that the equations of electrodynamics are the same for all observers. In particular, the vacuum velocity of light has the same value for all uniformly moving observers. This has implications for mechanics, especially for the relation between energy and momentum. Newtonian mechanics is a limiting case of relativistic dynamics, in the limit of velocities that are small compared to the velocity of light. *Relativistic quantum field theory* is a quantum field theory that is subject to the transformations of special relativity. The *theory of general relativity* is essentially a theory of gravitation that modifies our concepts of physical space and time even more profoundly than special relativity does. . . . . Sections 1.2, 1.4.1.

renormalization. Handling of the infinities of a quantum field theory. In a *renormalizable theory*, only a finite number of free parameters occur, and one may calculate probability amplitudes as a formal power series to any order. In QED the renormalized parameters are the charge and the mass of the electron. In an *unrenormalizable theory*, new parameters have to be introduced at each stage of refinement (each order of perturbation). . . . . Section 1.4.2.

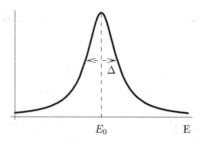

Figure A.2. A resonance curve at position E_0 with half-width Δ. For example, the cross section for scattering of A on B displayed versus the total energy of A and B.

representation of a group. Algebraic representation of an abstractly or geo-metrically defined group through matrices. The representation of the rotational group is by 3×3 matrices that preserve lengths and angles. ....Section 1.5.1.

resonance. An unstable state with a mean lifetime T, decaying into states A and B, leads to a resonance, that is, a peak in the cross section for scat-tering of A on B. A typical curve for a resonance at position E_0 with half-width Δ is displayed in Figure A.2. The half-width Δ is inversely propor-tional to the mean lifetime T: $\Delta = \hbar/T$. .... Section 2.5.

rest energy. The energy that is released if the mass of a particle is transformed into another form of energy. It is the product of the mass and the square of the speed of light in vacuum, $E_{rest} = mc^2$. .... Section 1.4.1.

SLAC. Stanford Linear Accelerator Center. ....Section 5.8.

S matrix. Probability for a scattering process. .... Section 3.1.

space reflection. Reflection of points in space around a preferred central point. Under space reflections, the Cartesian coordinates of a point change sign. .... Section 1.5.4.

spin. Property of elementary particles and nuclei. The concept is fully under-standable only in the mathematical framework of quantum mechanics. The unit for spin is the Planck constant $\hbar$. .... Sections 1.2, 1.5.2.

spin and statistics, theorem of. This theorem states that in a local quantum field theory, particles with integer spin are subject to Bose–Einstein statistics; particles with half-integer spin are subject to Fermi–Dirac statistics. .... Section 2.4.

spontaneous symmetry breaking. Breaking of a symmetry in which the field equations do not change, but the ground state (vacuum) is not invariant under the symmetry transformations. .... Section 5.3.

statistical error; standard error. An indicator of the range around the experimental mean value in which the "true" value of a quantity will lie with a probability of $\frac{2}{3}$. .... Section 1.4.3.

strangeness. A chargelike property of quarks and hadrons that is conserved in strong and electromagnetic interactions. *See also* **flavor**. .... Section 2.2.

SU(2); SU(3). Transformation groups for symmetries that play an important role in particle physics. SU(n) stands for special unitary matrices in n dimensions, that is, generalized rotations in a space with n dimensions and complex coordinates. .... Sections 1.5.2, 4.2.

symmetry. If certain properties are unchanged under certain transformations (*symmetry transformations*), one speaks of a symmetry under this transformation (rotation, for example). If the properties change, but only a little, one says that the symmetry is *broken* (from Greek *symmetria*, good proportion). .... Sections 1.5, 5.1–5.3.

synchrocyclotron. Cyclotron in which the laws of relativistic mechanics are taken into account. .... Section 1.7.

synchrotron. Particle accelerator, a further development of the synchrocyclotron. .... Section 1.7.

unitarity. Property of generalized rotational operators. The unitarity of the S matrix is a consequence of probability conservation (from Latin *unitas*, unity). .... Section 3.2.

vector meson. Meson with spin 1 and parity −1. .... Section 2.7.

B

Physical Units

All physical quantities can be expressed in units of length, time, and mass. The standard scientific units are the meter [m], the second [sec], and the kilogram [kg]. The fourth standard unit, the ampere [A], which is used for electric current, is measured in terms of forces and can therefore be expressed in units of length, time, and mass.

The standard units are fixed by convention, but there also exist *natural units*. Max Planck introduced them in 1899, employing fundamental constants of nature. An obvious natural unit for velocity is the velocity of light in vacuum, c, while the natural unit for action (energy times time) is the Planck constant, $\hbar$; it is also the natural unit for angular momentum. A natural unit for mass can be derived from Newton's gravitational constant G_N. This leads to the Planck mass, m_P, as the unit of mass. The values of the constants c, $\hbar$, and m_P, as well as the natural units of length, time, and mass, are given in Table B.1.

In particle physics, one employs the natural units $\hbar$ and c, but the Planck mass m_P is not in general use. Instead, the electron volt [eV] has been introduced as the unit of energy. It is the energy gained by an electron accelerated through a potential of one volt. In particle physics, the MeV, one million electron volts, and the GeV, one thousand MeV, are convenient units. Because of the famous relation between mass and energy $E = mc^2$, it is meaningful to express mass in GeV/c^2.

The unit of length in natural units is determined by the unit of energy. Nevertheless, in particle physics one generally employs a metric unit, the femtometer. One femtometer is a millionth of a nanometer, that is, 10^{-15} meters. The prefix *femto-* comes from Danish or Norwegian and means fifteen. The choice of this word was influenced by the original designation for 10^{-15} meters, the *fermi*, abbreviated fm, which could easily make the transition to stand for fem-

	symbol	value in given units		
		natural	standard	particle phys.
velocity of light in vacuum	c	1	$3 \cdot 10^8$ m/sec	
Planck constant	$\hbar$	1	10^{-34} kg·m^2/sec	$6.6 \cdot 10^{-25}$ GeV·sec
Newton gravitational constant	G_N	1	$6.7 \cdot 10^{-11}$ m^3/(kg·s^2)	
Planck mass	m_P	$\sqrt{\hbar \cdot c / G_N}$	$2.2 \cdot 10^{-8}$ kg	$1.2 \cdot 10^{19}$ GeV/c^2
elementary charge	e	$\sqrt{4\pi \cdot \hbar \cdot c / 137}$	$1.6 \cdot 10^{-19}$ A·sec	
Giga-electron volt	GeV	$8 \cdot 10^{-20}$ $m_P \cdot c^2$	$1.6 \cdot 10^{-10}$ kg·m^2/sec^2	5 $\hbar \cdot c$/fm
mass unit	GeV/c^2	$8 \cdot 10^{-20}$ m_P	$1.8 \cdot 10^{-27}$ kg	5 $\hbar$/(fm·c)
femtometer	fm	$6.6 \cdot 10^{19}$ $\hbar/(m_P \cdot c)$	10^{-15} m	5 $\hbar \cdot c$/GeV

Table B.1. Rounded numerical values and conversion factors for some important natural constants and units.

number	exponential representation	prefix	abbreviation	British number
trillion	10^{12}	tera-	T	billion
billion	10^9	giga-	G	thousand millions
million	10^6	mega-	M	
thousand	10^3	kilo-	k	
thousandth	10^{-3}	milli-	m	
millionth	10^{-6}	micro-	μ	
billionth	10^{-9}	nano-	n	
trillionth	10^{-12}	pico-	p	
quadrillionth	10^{-15}	femto-	f	

Table B.2. Prefixes and powers of ten.

tometer. One can express the unit fm in units of GeV through the relation $1\,\text{GeV} \cdot 1\,\text{fm} = 5.08\,\hbar \cdot c$. The use of separate units for energy and length is unnatural in a system of natural units, and the irrationality of it—such is present in all professions—becomes evident through the introduction of a specific unit for surface area. This unit is the *barn*. It is not as big as a barn, though it seems so in the world of nuclear physics. One barn equals $100\,\text{fm}^2$.

Charges are expressed in units of the elementary charge e, that is, in units of the charge of a proton.

Also useful, although not very profound, is Table B.2; it contains the prefixes and associated factors of powers of ten used with units.

C

Nobel Prize Winners

The following are the Nobel Prize winners mentioned in this book.

Alvarez, Luis Walter. *San Francisco 6/13/1911, †Berkeley 9/1/1988.
Physics prize, 1968, "for his decisive contributions to elementary particle physics, in particular the discovery of a large number of resonance states, made possible through his development of the technique of using hydrogen bubble chamber and data analysis."

Anderson, Carl David. *New York 9/3/1905, †San Marino 1/11/1991.
Physics prize, 1936, "for his discovery of the positron."

Becquerel, Antoine-Henri. *Paris 12/15/1852, †Le Croisic 8/25/1908.
Physics prize, 1903, "in recognition of the extraordinary services he has rendered by his discovery of spontaneous radioactivity."

Bethe, Hans Albrecht. *Strasbourg 7/2/1907, †Ithaca 3/6/2005.
Physics prize, 1967, "for his contributions to the theory of nuclear reactions, especially his discoveries concerning the energy production in stars."

Blackett, Patrick Maunard Stuart. *London 11/18.1897, †London 7/13/1974.
Physics prize, 1948, "for his development of the Wilson cloud chamber method, and his discoveries therewith in the fields of nuclear physics and cosmic radiation."

Bohr, Niels Henrik David. *Copenhagen 10/7/1885, †Copenhagen 11/18/1962.
Physics prize, 1922, "for his services in the investigation of the structure of atoms and of the radiation emanating from them."

Born, Max. *Breslau 12/11/1882, †Göttingen 1/5/1970.
Physics prize, 1954, "for his fundamental research in quantum mechanics, especially for his statistical interpretation of the wavefunction."

Bothe, Walter. *Oranienburg 1/8/1891, †Heidelberg 2/8/1957.
Physics prize, 1954, "for the coincidence method and his discoveries made therewith."

Chadwick, Sir James. *Manchester 10/20/1891, †Cambridge 7/24/1974.
Physics prize, 1935, "for the discovery of the neutron."

Chamberlain, Owen. *San Francisco 7/10/1920.
Physics prize, 1959, jointly with G. Segrè "for their discovery of the antiproton."

Charpak, Georges. *Dabrowiza 8/1/1924.
Physics prize, 1992, "for his invention and development of particle detectors, in particular the multiwire proportional chamber."

Cherenkov, Pavel Alekseyevich. *Voronezh 7/28/1904, †Moscow 1/6/1990.
Physics prize, 1958, jointly with I. M. Frank and I. Y. Tamm "for the discovery and the interpretation of the Cherenkov effect."

Compton, Arthur Holly. *Wooster 9/10/1892, †Berkeley 3/15/1962.
Physics prize, 1927, "for his discovery of the effect named after him."

Cockcroft, Sir John Douglas. *Todmorden 5/27/1897, †Cambridge 9/18/1967.
Physics prize, 1951, jointly with E. T. S. Walton "for their pioneer work on the transmutation of atomic nuclei by artificially accelerated atomic particles."

Cronin, James Watson. *Chicago 9/29/1931.
Physics prize, 1980, jointly with V. L. Fitch "for the discovery of violations of fundamental symmetry principles in the decay of neutral K-mesons."

Curie, Marie, née Sklodowska. *Warsaw 11/7/1867, †Sancellemoz 7/4/1934.
Physics prize, 1903, jointly with P. Curie "in recognition of the extraordinary services they have rendered by their joint researches on the radiation phenomena discovered by Professor Henri Becquerel."

Curie, Pierre. *Paris 5/15/1859, †Paris 4/19/1906.

Physics prize, 1903, jointly with M. Curie "in recognition of the extraordinary services they have rendered by their joint researches on the radiation phenomena discovered by Professor Henri Becquerel."

Davis, Raymond, Jr. *Washington DC 10/14/1914.

Physics prize, 2002, jointly with M. Koshiba "for pioneering contributions to astrophysics, in particular for the detection of cosmic neutrinos."

de Broglie, Louis-Victor Raymond. *Dieppe 8/15/1892, †Louveciennes 3/19/1987.

Physics prize, 1929, "for his discovery of the wave nature of electrons."

Dehmelt, Hans Georg. *Görlitz 9/9/1922.

Physics prize, 1989, jointly with W. Paul "for the development of the ion trap technique."

Dirac, Paul Adrien Maurice. *Bristol 8/8/1902, †Tallahassee 10/20/1984.

Physics prize, 1933, jointly with E. Schrödinger "for the discovery of new productive forms of atomic theory."

Einstein, Albert. *Ulm 3/14/1879, †Princeton 4/18/1955.

Physics prize, 1921, "for his services to Theoretical Physics, and especially for his discovery of the law of the photoelectric effect."

Fermi, Enrico. *Rome 9/29/1901, †Chicago 11/28/1954.

Physics prize, 1938, "for his demonstrations of the existence of new radioactive elements produced by neutron irradiation, and for his related discovery of nuclear reactions brought about by slow neutrons."

Feynman, Richard. *New York 5/11/1918, †Los Angeles 2/15/1988.

Physics prize, 1965, jointly with J. Schwinger and S. Tomonaga "for their fundamental work in quantum electrodynamics, with deep-ploughing consequences for the physics of elementary particles."

Fitch, Val Logsdon. *Merriman 3/10/1923.

Physics prize, 1980, jointly with J. W. Cronin "for the discovery of violations of fundamental symmetry principles in the decay of neutral K-mesons."

Franck, James. *Hamburg 8/26/1882, †Göttingen 5/21/1964.

Physics prize, 1925, jointly with G. Hertz "for their discovery of the laws governing the impact of an electron upon an atom."

Frank, Ilya Mikhailovich. *St. Petersburg 10/23/1908, †Moscow 6/22/1990.
Physics prize, 1958, jointly with P. A. Cherenkov and I. Y. Tamm "for the discovery and the interpretation of the Cherenkov effect."

Friedman, Jerome Isaac. *Chicago 3/28/1930.
Physics prize, 1990, jointly with H. W. Kendall and R. E. Taylor "for their pioneering investigations concerning deep inelastic scattering of electrons on protons and bound neutrons, which have been of essential importance for the development of the quark model in particle physics."

Gell-Mann, Murray. *New York 9/15/1929.
Physics prize, 1969, "for his contributions and discoveries concerning the classification of elementary particles and their interactions."

Glaser, Donald Arthur. *Cleveland 9/21/1926.
Physics prize, 1960, "for the invention of the bubble chamber."

Glashow, Sheldon Lee. *New York 12/5/1932.
Physics prize, 1979, jointly with A. Salam and S. Weinberg "for their contributions to the theory of the unified weak and electromagnetic interaction between elementary particles, including, inter alia, the prediction of the weak neutral current."

Goeppert-Mayer, Maria. *Katowice 6/28/1906, †San Diego 2/20/1972.
Physics prize, 1963, jointly with J. H. D. Jensen "for their discoveries concerning nuclear shell structure."

Gross, David J. *Washington, DC 2/19/1941.
Physics prize, 2004, jointly with H. D. Politzer and F. Wilczek, "for the discovery of asymptotic freedom in the theory of the strong interaction."

Heisenberg, Werner Karl. *Würzburg 12/5/1901, †Munich 2/1/1976.
Physics prize, 1932, "for the creation of quantum mechanics, the application of which has, inter alia, led to the discovery of the allotropic forms of hydrogen."

Hertz, Gustav. *Hamburg 7/22/1887, †Berlin 10/30/1974.
Physics prize, 1925, jointly with J. Franck "for their discovery of the laws governing the impact of an electron upon an atom."

Hess, Viktor Franz. *Schloss Waldstein 6/3/1883, †Mount Vernon 12/17/1964.
Physics prize, 1936, "for his discovery of cosmic radiation."

Hofstadter, Robert. *New York 2/5/1915, †Stanford 11/17/1990.

Physics prize, 1961, "for his pioneering studies of electron scattering in atomic nuclei and for his thereby achieved discoveries concerning the structure of the nucleons."

Jensen, Johannes Daniel. *Hamburg 6/25/1907, †Heidelberg 2/11/1973.

Physics prize, 1963, jointly with M. Goeppert-Mayer "for their discoveries concerning nuclear shell structure."

Joliot-Curie, Irène. *Paris 8/12/1897, †Paris 3/16/1956.

Chemistry prize, 1935, jointly with F. Joliot "in recognition of their synthesis of new radioactive elements."

Joliot, Frédéric. *Paris 3/12/1900, †Paris 8/14/1959.

Chemistry prize, 1935, jointly with I. Joliot-Curie "in recognition of their synthesis of new radioactive elements."

Kendall, Henry Way. *Boston 12/9/1926.

Physics prize, 1990, jointly with J. I. Friedman and R. E. Taylor "for their pioneering investigations concerning deep inelastic scattering of electrons on protons and bound neutrons, which have been of essential importance for the development of the quark model in particle physics."

Koshiba, Masatoshi. *Toyohashi 9/19/1926.

Physics prize, 2002, jointly with R. Davis "for pioneering contributions to astrophysics, in particular for the detection of cosmic neutrinos."

Kusch, Polykarp. *Blankenburg 1/26/1911, † Dallas 3/20/1993.

Physics prize, 1955, "for his precision determination of the magnetic moment of the electron."

Landau, Lev Davidovich. *Baku 1/22/1908, †Moscow 4/1/1968.

Physics prize, 1962, "for his pioneering theories for condensed matter, especially liquid helium."

Lawrence, Ernest Orlando. *Canton 8/8/1901, †Palo Alto 8/27/1958.

Physics prize, 1939, "for the invention and development of the cyclotron and for results obtained with it, especially with regard to artificial radioactive elements."

Lederman, Leon Max. *New York 7/15/1922.

Physics prize, 1988, jointly with M. Schwartz and J. Steinberger "for the neutrino beam method and the demonstration of the doublet structure of the leptons through the discovery of the muon neutrino."

Lee, Tsung Dao. *Shanghai 11/25/1926.

Physics prize, 1957, jointly with C. N. Yang "for their penetrating investigation of the so-called parity laws which has led to important discoveries regarding the elementary particles."

Lenard, Philipp Eduard Anton. *Pressburg 6/7/1862, †Messelhausen 5/20/1947.

Physics prize, 1905, "for his work on cathode rays."

Lorentz, Hendrik Antoon. *Arnheim 7/18/1853, †Haarlem 2/4/1928.

Physics prize, 1902, jointly with P. Zeeman "in recognition of the extraordinary service they rendered by their researches into the influence of magnetism upon radiation phenomena."

Millikan, Robert Andrews. *Morrison 3/22/1868, †Pasadena 12/19/1953.

Physics prize, 1923, "for his work on the elementary charge of electricity and on the photoelectric effect."

Paul, Wolfgang. *Lorenzkirch 8/10.1913, †Bonn 12/7/1993.

Physics prize, 1989, jointly with H. G. Dehmelt "for the development of the ion trap technique."

Pauli, Wolfgang Ernst Friedrich. *Vienna 4/25/1900, †Zürich 12/15/1958.

Physics prize, 1945, "for the discovery of the Exclusion Principle, also called the Pauli Principle."

Penzias, Arno Allen. *Munich 4/26/1933.

Physics prize, 1978, jointly with R. W. Wilson "for their discovery of cosmic microwave background radiation."

Perl, Martin Lewis. *New York 6/24/1927.

Physics prize, 1995, "for pioneering experimental contributions to lepton physics, especially for the discovery of the tau lepton."

Planck, Max. *Kiel 4/23/1858, †Göttingen 10/4/1947.

Physics prize, 1918, "in recognition of the services he rendered to the advancement of Physics by his discovery of energy quanta."

Politzer, H. David. * New York, NY 8/31/1949.

Physics prize, 2004, jointly with D. J. Gross and F. Wilczek, "for the discovery of asymptotic freedom in the theory of the strong interaction."

Powell, Cecil Frank. * Tonbridge 12/5/1903, † Bellano 8/9/1969.

Physics prize, 1950, "for his development of the photographic method of studying nuclear processes and his discoveries regarding mesons made with this method."

Rabi, Isaac Isidor. * Rymanóv 7/29/1898, † New York 1/11/1988.

Physics prize, 1944, "for his resonance method for recording the magnetic properties of atomic nuclei."

Reines, Frederick. * Paterson 3/16/1918, † Irvine 8/26/1998.

Physics prize, 1995, "for pioneering experimental contributions to lepton physics, especially for the detection of the neutrino."

Richter, Burton. * New York 3/22/1931.

Physics prize, 1976, jointly with S. C. C. Ting "for their pioneering work in the discovery of a heavy elementary particle of a new kind."

Röntgen, Wilhelm Conrad. * Lennep 3/27/1845, † Munich 2/10/1923.

Physics prize, 1901, "in recognition of the extraordinary services he has rendered by the discovery of the remarkable rays subsequently named after him."

Rubbia, Carlo. * Gorizia 3/31/1934.

Physics prize, 1984, jointly with S. van der Meer "for their decisive contributions to the large project, which led to the discovery of the field particles W and Z, communicators of weak interaction."

Rutherford, Ernest. * Nelson 8/30/1871, † Cambridge 10/19/1937.

Chemistry prize, 1908, "for his investigations into the disintegration of the elements, and the chemistry of radioactive substances."

Salam, Abdus. * Jhang 1/29/1926, † Oxford 11/21/1996.

Physics prize, 1979, jointly with S. L. Glashow and S. Weinberg "for their contributions to the theory of the unified weak and electromagnetic interaction between elementary particles, including, inter alia, the prediction of the weak neutral current."

Schrödinger, Erwin. *Vienna 12/18/1987, †Vienna, 1/4/1961.
Physics prize, 1933, jointly with P. A. M. Dirac "for the discovery of new productive forms of atomic theory."

Schwartz, Melvin. *New York 11/2/1932.
Physics prize, 1988, jointly with L. M. Lederman and J. Steinberger "for the neutrino beam method and the demonstration of the doublet structure of the leptons through the discovery of the muon neutrino."

Schwinger, Julian. *New York 2/12/1918, †Los Angeles 7/16/1994.
Physics prize, 1965, jointly with R. Feynman and S. Tomonaga "for their fundamental work in quantum electrodynamics, with deep-ploughing consequences for the physics of elementary particles."

Segrè, Emilio Gino. *Tivoli 2/1/1905, †Lafayette 4/22/1989.
Physics prize, 1959, jointly with O. Chamberlain "for their discovery of the antiproton."

Steinberger, Jack. *Bad Kissingen 5/25/1921.
Physics prize, 1988, jointly with L. M. Lederman and M. Schwartz "for the neutrino beam method and the demonstration of the doublet structure of the leptons through the discovery of the muon neutrino."

Stern, Otto. *Sorau 2/17/1888, †Berkeley 8/17/1969.
Physics prize, 1943, "for his contribution to the development of the molecular ray method and his discovery of the magnetic moment of the proton."

Tamm, Igor Evgenevich. *Vladivostok 7/8/1895, †Moscow 4/12/1971.
Physics prize, 1958, jointly with P. A. Cherenkov and I. Y. Frank "for the discovery and the interpretation of the Cherenkov effect."

Taylor, Richard Edward. *Medicine Hat 11/2/1929.
Physics prize, 1990, jointly with J. I. Friedman and H. W. Kendall "for their pioneering investigations concerning deep inelastic scattering of electrons on protons and bound neutrons, which have been of essential importance for the development of the quark model in particle physics."

Thomson, Sir Joseph John. *Cheetham Hill 12/18/1856, †Cambridge 8/30/1940.
Physics prize, 1906, "in recognition of the great merits of his theoretical and experimental investigations on the conduction of electricity by gases."

't Hooft, Gerardus. *Den Helder 7/5/1946.
Physics prize, 1999, jointly with M. J. G. Veltman "for elucidating the quantum structure of electroweak interactions in physics."

Ting, Samuel Chao Chung. *Ann Arbor 1/27/1936.
Physics prize, 1976, jointly with B. Richter "for their pioneering work in the discovery of a heavy elementary particle of a new kind."

Tomonaga, Shin-Ichiro. *Tokyo 3/31/1906, †Tokyo 7/8/1979.
Physics prize, 1965, jointly with R. Feynman and J. Schwinger "for their fundamental work in quantum electrodynamics, with deep-ploughing consequences for the physics of elementary particles."

van der Meer, Simon. *Den Haag 11/24/1925.
Physics prize, 1984, jointly with C. Rubbia "for their decisive contributions to the large project, which led to the discovery of the field particles W and Z, communicators of weak interaction."

Veltman, Martinus. *Waalwijk 6/27/1931.
Physics prize, 1999, jointly with G. 't Hooft "for elucidating the quantum structure of electroweak interactions in physics."

Walton, Ernest Thomas Sinton. *Dungarvan 10/6/1903, †Belfast 6/25/1995.
Physics prize, 1951, jointly with J. D. Cockcroft "for their pioneer work on the transmutation of atomic nuclei by artificially accelerated atomic particles."

Weinberg, Steven. *New York 5/3/1933.
Physics prize, 1979, jointly with S. L. Glashow and A. Salam "for their contributions to the theory of the unified weak and electromagnetic interaction between elementary particles, including, inter alia, the prediction of the weak neutral current."

Wigner, Eugene Paul. *Budapest 11/17/1902, †Princeton 1/1/1995.
Physics prize, 1963, "for his contributions to the theory of the atomic nucleus and the elementary particles, particularly through the discovery and application of fundamental symmetry principles."

Wilczek, Frank. *New York, NY 5/15/1951.
Physics prize, 2004, jointly with D. J. Gross and H. D. Politzer, "for the discovery of asymptotic freedom in the theory of the strong interaction."

Wilson, Charles Thomson Rees. *Glencorse 2/14/1869, †Carlops 11/15/1959.
Physics prize, 1927, "for his method of making the paths of electrically charged particles visible by condensation of vapor."

Wilson, Kenneth Geddes. *Waltham 6/8/1936.
Physics prize, 1982, "for his theory for critical phenomena in connection with phase transitions."

Wilson, Robert Woodrow. *Houston 1/10/1936.
Physics prize, 1978, jointly with A. A. Penzias "for their discovery of cosmic microwave background radiation."

Yang, Chen Ning. *Hefei 9/22/1922.
Physics prize, 1957, jointly with T. D. Lee "for their penetrating investigation of the so-called parity laws which has led to important discoveries regarding the elementary particles."

Yukawa, Hideki. *Tokyo 1/23/1907, †Kyoto 9/8/1981.
Physics prize, 1949, "for his prediction of the existence of mesons on the basis of theoretical work on nuclear forces."

D

Recommended Reading

This is a very personal and incomplete selection of books and conference reports that may be of interest to the reader.

I have mentioned several times in the text the beautiful book by Abraham Pais, *Inward Bound: Of Matter and Forces in the Physical World*, Oxford University Press, 1986.

I found the following conference reports dealing with the history of particle physics particularly interesting:

- *International Colloquium on the History of Particle Physics, Journal de Physique*, Volume 43 (1982) C-8, especially the contributions of C. Peyrou, H. L. Anderson and N. Kemmer.

- *The Rise of the Standard Model: A History of Particle Physics from 1964 to 1979*, edited by Lillian Hoddeson et al., Cambridge University Press, 1997, especially the contributions of L. Lederman, M. Veltman, R. Schwitters, S. Bludman, J. Iliopoulos, J. Bjorken, S. L. Wu, and M. Gell-Mann.

Scientific American also regularly publishes articles on special subjects of particle physics.

There are many textbooks in the area; I list here only a few (in ascending order of difficulty):

- *Particles and Nuclei: An Introduction to the Physical Concepts*, 5th edition, Bogdan Povh, Klaus Rith, Christoph Scholz, and Frank Zetsche (translated from the German by Martin Lavelle), Springer, 2006.

- *Elementary Particle Physics: Concepts and Phenomena*, Texts and Monographs in Physics, O. Nachtmann (translated from the German by W. Wetzel), Springer 1990.

- *An Introduction to Quantum Field Theory*, Frontiers in Physics, Michael E. Peskin and Dan V. Schroeder, HarperCollins Publishers, 1995.

I also recommend some books written by the principal players in the field:

- *QED: The Strange Theory of Light and Matter*, Princeton Science Library, Richard P. Feynman, Princeton University Press, 2006.

- *Thirty Years That Shook Physics: The Story of Quantum Theory*, George Gamow, Dover, 1985.

- *The Quark and the Jaguar: Adventures in the Simple and Complex*, Murray Gell-Mann, W. H. Freeman & Company, 1994.

- *Learning About Particles — 50 Priveleged Years*, Jack Steinberger, Springer, 2004.

- *Facts and Mysteries in Elementary Particle Physics*, Martinus Veltman, World Scientific Publishing Company, 2003.

- *The Joy of Insight: Passions of a Physicist*, Alfred P. Sloan Foundation Series, Victor Weisskopf, Basic Books, 1991.

A listing of a large selection of the original literature published in professional journals as well as additional books on specific topics can be found on the book's website (http://www.thphys.uni-heidelberg.de/~dosch/beyondnano).

Index

279